CHEVY II NOVA & SS
1962 - 1972

Compiled by
R.M. Clarke

ISBN 0 948207 29 9

Distributed by
Brooklands Book Distribution Ltd.
'Holmerise', Seven Hills Road,
Cobham, Surrey, England
Printed in Hong Kong

BROOKLANDS BOOKS

BROOKLANDS BOOKS SERIES
AC Ace & Aceca 1953-1983
AC Cobra 1962-1969
Alfa Romeo Alfasud 1972-1984
Alfa Romeo Alfetta Coupes GT.GTV.GTV6 1974-1987
Alfa Romeo Guilias Berlinettas
Alfa Romeo Giulia Berlinas 1962-1976
Alfa Romeo Giulia Coupés 1963-1976
Alfa Romeo Spider 1966-1987
Allard Gold Portfolio 1937-1958
Aston Martin Gold Portfolio 1972-1985
Austin Seven 1922-1982
Austin A30 & A35 1951-1962
Austin Healey 100 1952-1959
Austin Healey 3000 1959-1967
Austin Healey 100 & 3000 Collection No. 1
Austin Healey 'Frogeye' Sprite Collection No. 1
Austin Healey Sprite 1958-1971
Avanti 1962-1983
BMW Six Cylinder Coupés 1969-1975
BMW 1600 Collection No. 1
BMW 2002 1968-1976
Bristol Cars Gold Portfolio 1946-1985
Buick Automobiles 1947-1960
Buick Riviera 1963-1978
Cadillac Automobiles 1949-1959
Cadillac Automobiles 1960-1969
Cadillac Eldorado 1967-1978
Camaro 1966-1970
Chevrolet Camaro & Z-28 1973-1981
High Performance Camaros 1982-1988
Chevrolet Camaro Collection No. 1
Chevrolet 1955-1957
Chevrolet Impala & SS 1958-1971
Chevelle & SS 1964-1972
Chevy II Nova & SS 1962-1973
Chrysler 300 1955-1970
Citroen Traction Avant 1934-1957
Citroen DS & ID 1955-1875
Citroen 2CV 1948-1988
Cobras & Replicas 1962-1983
Cortina 1600E & GT 1967-1970
Corvair 1959-1968
Daimler Dart & V-8 250 1959-1969
Datsun 240z 1970-1973
Datsun 280Z & ZX 1975-1983
De Tomaso Collection No. 1
Dodge Charger 1966-1974
Excalibur Collection No. 1
Ferrari Cars 1946-1956
Ferrari Cars 1962-1966
Ferrari Cars 1969-1973
Ferrari Dino 1965-1974
Ferrari Dino 308 1974-1979
Ferrari 308 & Mondial 1980-1984
Ferrari Collection No. 1
Fiat-Bertone X1/9 1973-1988
Fiat Pininfarina 124+2000 Spider 1968-1985
Ford Falcon 1960-1970
Ford Mustang 1964-1967
Ford Mustang 1967-1973
High Performance Mustangs 1982-1988
Ford RS Escort 1968-1980
Honda CRX 1983-1987
High Performance Escorts MkI 1968-1974
High Performance Escorts MkII 1975-1980
Hudson & Railton Cars 1936-1940
Jaguar XK120 XK140 XK150 Gold Portfolio 1948-1960
Jaguar Cars 1957-1961
Jaguar Cars 1961-1964
Jaguar MK2 1959-1969
Jaguar E-Type 1961-1966
Jaguar E-Type 1966-1971
Jaguar E-Type V12 1971-1975
Jaguar XKE Collection No. 1
Jaguar XJ6 1968-1972
Jaguar XJ6 Series II 1973-1979
Jaguar XJ6 & XJ12 Series III 1979-1985
Jaguar XJ12 1972-1980
Jaguar XJS Gold Portfolio 1975-1988
Jensen Cars 1946-1967
Jensen Cars 1967-1979
Jensen Interceptor Gold Portfolio 1966-1986
Lamborghini Cars 1964-1970
Lamborghini Cars 1970-1975
Lamborghini Countach Collection No. 1
Lamborghini Countach & Urraco 1974-1980
Lamborghini Countach & Jalpa 1980-1985
Lancia Stratos 1972-1985
Land Rover 1948-1973
Land Rover Series II & IIa 1958-1971
Land Rover Series III 1971-1985
Land Rover 90 & 110 1983-1989
Lotus Cortina 1963-1970
Lotur Elan Gold Portfolio 1962-1974
Lotus Elan Collection No. 2
Lotus Elite 1957-1964
Lotus Elite & Eclat 1974-1981
Lotus Turbo Esprit 1980-1986
Lotus Europa 1966-1975
Lotus Europa Collection No. 1
Lotus Seven 1957-1980
Lotus Seven Collection No. 1
Maserati 1965-1970
Maserati 1970-1975
Marcos Cars 1960-1988
Mazda RX-7 Collection No. 1
Mercedes 190 & 300SL 1954-1963
Mercedes 230/250/280SL 1963-1971
Mercedes 350/450SL & SLC 1971-1980
Mercedes Benz Cars 1949-1954
Mercedes Benz Cars 1954-1957
Mercedes Benz Cars 1957-1961
Mercedes Benz Competition Cars 1950-1957
Metropolitan 1954-1962
MG Cars 1929-1934
MG TC 1945-1949
MG TD 1949-1953
MG TF 1953-1955
MG Cars 1957-1959
MG Cars 1959-1962
MG Midget 1961-1980
MGA Collection No. 1
MGA Roadsters 1955-1962
MGB Roadsters 1962-1980
MGB GT 1965-1980
Mini Cooper 1961-1971
Morgan Cars 1960-1970
Morgan Cars 1969-1979
Morris Minor Collection No. 1
Old's Cutlass & 4-4-2 1964-1972
Oldsmobile Toronado 1966-1978
Opel GT 1968-1973
Packard Gold Portfolio 1946-1958
Pantera 1970-1973
Pantera & Mangusta 1969-1974
Plymouth Barracuda 1964-1974
Pontiac Fiero 1984-1988
Pontiac GTO 1964-1970
Pontiac Firebird 1967-1973
Pontiac Firebird and Trans-Am 1973-1981
High Performance Firebirds 1982-1988
Pontiac Tempest & GTO 1961-1965
Porsche Cars 1960-1964
Porsche Cars 1964-1968
Porsche Cars 1968-1972
Porsche Cars in the Sixties
Porsche Cars 1972-1975
Porsche 356 1952-1965
Porsche 911 Collection No. 1
Porsche 911 Collection No. 2
Porsche 911 1965-1969
Porsche 911 1970-1972
Porsche 911 1973-1977
Porsche 911 Carrera 1973-1977
Porsche 911 SC 1978-1983
Porsche 911 Turbo 1975-1984
Porsche 914 Gold Portfolio 1969-1988
Porsche 914 Collection No. 1
Porsche 924 1975-1981
Porsche 928 Collection No. 1
Porsche 944 1981-1985
Porsche Turbo Collection No. 1
Reliant Scimitar 1964-1986
Riley 1½ & 2½ Litre Gold Portfolio 1945-1955
Rolls Royce Silver Cloud 1955-1965
Rolls Royce Silver Shadow 1965-1980
Range Rover Gold Portfolio 1970-1988
Rover 3 & 3.5 Litre 1958-1973
Rover P4 1949-1959
Rover P4 1955-1964
Rover 2000 + 2200 1963-1977
Rover 3500 1968-1976
Rover 3500 & Vitesse 1976-1986
Saab Sonett Collection No. 1
Saab Turbo 1976-1983
Studebaker Hawks & Larks 1956-1963
Sunbeam Tiger And Alpine Gold Portfolio 1959-1967
Thunderbird 1955-1957
Thunderbird 1958-1963
Thunderbird 1964-1976
Toyota MR2 1984-1988
Triumph 2000-2.5-2500 1963-1977
Triumph Spitfire 1962-1980
Triumph Spitfire Collection No. 1
Triumph Stag 1970-1980
Triumph Stag Collection No. 1
Triumph TR2 & TR3 1952-1960
Triumph TR4.TR5.TR250 1961-1968
Triumph TR6 1969-1976
Triumph TR6 Collection No. 1
Triumph TR7 & TR8 1975-1982
Triumph GT6 1966-1974
Triumph Vitesse & Herald 1959-1971
TVR Gold Portfolio 1959-1988
Volkswagen Cars 1936-1956
VW Beetle 1956-1977
VW Beetle Collection No. 1
VW Golf GTi 1976-1986
VW Karmann Ghia 1955-1982
VW Scirocco 1974-1981
VW Bus-Camper-Van 1954-1967
VW Bus-Camper-Van 1968-1979
Volvo 1800 1960-1973
Volvo 120 Series 1956-1970

BROOKLANDS MUSCLE CARS SERIES
American Motors Muscle Cars 1966-1970
Buick Muscle Cars 1965-1970
Camaro Muscle Cars 1966-1972
Capri Muscle Cars 1965-1970
Chevrolet Muscle Cars 1966-1972
Dodge Muscle Cars 1967-1970
Mercury Muscle Cars 1966-1971
Mini Muscle Cars 1961-1979
Mopar Muscle Cars 1964-1967
Mopar Muscle Cars 1968-1971
Mustang Muscle Cars 1967-1971
Shelby Mustang Muscle Cars 1965-1970
Oldsmobile Muscle Cars 1964-1970
Plymouth Muscle Cars 1966-1971
Pontiac Muscle Cars 1966-1972
Muscle Cars Compared 1966-1971
Muscle Cars Compared Book 2 1965-1971

BROOKLANDS ROAD & TRACK SERIES
Road & Track on Alfa Romeo 1949-1963
Road & Track on Alfa Romeo 1964-1970
Road & Track on Alfa Romeo 1971-1976
Road & Track on Alfa Romeo 1977-1984
Road & Track on Aston Martin 1962-1984
Road & Track on Auburn Cord & Duesenberg 1952-1984
Road & Track on Audi 1952-1980
Road & Track on Audi 1980-1986
Road & Track on Austin Healey 1953-1970
Road & Track on BMW Cars 1966-1974
Road & Track on BMW Cars 1975-1978
Road & Track on BMW Cars 1979-1983
Road & Track on Cobra, Shelby & Ford GT40 1962-1983
Road & Track on Corvette 1953-1967
Road & Track on Corvette 1968-1982
Road & Track on Corvette 1982-1986
Road & Track on Datsun Z 1970-1983
Road & Track on Ferrari 1950-1968
Road & Track on Ferrari 1968-1974
Road & Track on Ferrari 1975-1981
Road & Track on Ferrari 1981-1984
Road & Track on Fiat Sports Cars 1968-1987
Road & Track on Jaguar 1950-1960
Road & Track on Jaguar 1961-1968
Road & Track on Jaguar 1968-1974
Road & Track on Jaguar 1974-1982
Road & Track on Jaguar 1983-1989
Road & Track on Lamborghini 1964-1985
Road & Track on Lotus 1972-1981
Road & Track on Maserati 1952-1974
Road & Track on Maserati 1975-1983
Road & Track on Mazda RX7 1978-1986
Road & Track on Mercedes 1952-1962
Road & Track on Mercedes 1963-1970
Road & Track on Mercedes 1971-1979
Road & Track on Mercedes 1980-1987
Road & Track on MG Sports Cars 1949-1961
Road & Track on MG Sports Cars 1962-1980
Road & Track on Mustang 1964-1977
Road & Track on Peugeot 1955-1986
Road & Track on Pontiac 1960-1983
Road & Track on Porsche 1951-1967
Road & Track on Porsche 1968-1971
Road & Track on Porsche 1972-1975
Road & Track on Porsche 1975-1978
Road & Track on Porsche 1979-1982
Road & Track on Porsche 1982-1985
Road & Track on Rolls Royce & Bentley 1950-1965
Road & Track on Rolls Royce & Bentley 1966-1984
Road & Track on Saab 1955-1985
Road & Track on Toyota Sports & G T Cars 1966-1986
Road & Track on Triumph Sports Cars 1953-1967
Road & Track on Triumph Sports Cars 1967-1974
Road & Track on Triumph Sports Cars 1974-1982
Road & Track on Volkswagen 1951-1968
Road & Track on Volkswagen 1968-1978
Road & Track on Volkswagen 1978-1985
Road & Track on Volvo 1957-1974
Road & Track on Volvo 1975-1985
Road & Track Henry Manney At Large & Abroad

BROOKLANDS CAR AND DRIVER SERIES
Car and Driver on BMW 1955-1977
Car and Driver on BMW 1977-1985
Car and Driver on Cobra, Shelby & Ford GT40 1963-1984
Car and Driver on Datsun Z 1600 & 2000 1966-1984
Car and Driver on Corvette 1956-1967
Car and Driver on Corvette 1968-1977
Car and Driver on Corvette 1978-1982
Car and Driver on Corvette 1983-1988
Car and Driver on Ferrari 1955-1962
Car and Driver on Ferrari 1963-1975
Car and Driver on Ferrari 1976-1983
Car and Driver on Mopar 1956-1967
Car and Driver on Mopar 1968-1975
Car and Driver on Mustang 1964-1972
Car and Driver on Pontiac 1961-1975
Car and Driver on Porsche 1955-1962
Car and Driver on Porsche 1963-1970
Car and Driver on Porsche 1970-1976
Car and Driver on Porsche 1977-1981
Car and Driver on Porsche 1982-1986
Car and Driver on Saab 1956-1985
Car and Driver on Volvo 1955-1986

BROOKLANDS MOTOR & THOROUGHBRED & CLASSIC CAR SERIES
Motor & T & CC on Ferrari 1966-1976
Motor & T & CC on Ferrari 1976-1984
Motor & T & CC on Lotus 1979-1983
Motor & T & CC on Morris Minor 1948-1983

BROOKLANDS PRACTICAL CLASSICS SERIES
Practical Classics on Austin A 40 Restoration
Practical Classics on Land Rover Restoration
Practical Classics on Metalworking in Restoration
Practical Classics on Midget/Sprite Restoration
Practical Classics on Mini Cooper Restoration
Practical Classics on MGB Restoration
Practical Classics on Morris Minor Restoration
Practical Classics on Triumph Herald/Vitesse
Practical Classics on Triumph Spitfire Restoration
Practical Classics on VW Beetle Restoration
Practical Classics on 1930S Car Restoration

BROOKLANDS MILITARY VEHICLES SERIES
Allied Military Vehicles Collection No. 1
Allied Military Vehicles Collection No. 2
Dodge Military Vehicles Collection No. 1
Military Jeeps 1941-1945
Off Road Jeeps 1944-1971
V W Kubelwagen 1940-1975

BROOKLANDS BOOKS

CONTENTS

5	Driving the Chevy II	*Motor Trend*	Dec.	1961
7	Chevy II An Honourable Compromise Road Test	*Canada Track & Traffic*	Dec.	1961
10	Chevy II Road Test	*Car Life*	Feb.	1962
16	Chevrolet's Compact Chevy II Road Test	*Wheels*	July	1962
20	Nova SS Chevy II Sport Coupé Road Test	*Motor Trend*	May	1963
27	Chevy II 400 Nova Series 4-door	*World Car Catalogue*		1964
28	Chevy II V-8 Road Test	*Motor Trend*	June	1964
34	Chevy II 327/350 V-8 Road Test	*Car Life*	May	1966
38	Chevy II Nova 327 Road Test	*Motor Trend*	July	1966
42	Chevy II 1968 Report	*Motorcade*	Oct.	1967
44	Chevy's Pop Compact Road Test	*Hot Rod*	May	1968
47	The Chevy II that Options Built Road Test	*Car Life*	May	1968
52	Chevy II vs VW Comparison	*Motor Trend*	June	1968
57	Chevy II Nova SS Road Test	*Car and Driver*	Aug.	1968
62	Phase III SS427 Chevy Nova Road Test	*Cars*	Mar.	1969
68	Chevy's Secret Weapon the Nova SS	*Road Test*	May	1969
74	13-Second Grocery Getter Road Test	*Hot Rod*	July	1969
77	Nova SS350 Road Test	*Car Life*	Jan.	1970
82	Chevrolet Nova 396SS Coupé	*Road Test*	July	1970
88	The Seven Year Car Comparison Test	*Motor Trend*	Sept.	1971
94	Super Nova	*Car Craft*	Mar.	1972
97	A Compact with Pony Road Test	*Road Test*	Feb.	1972

ACKNOWLEDGEMENTS

The first Brooklands book made its appearance about 25 years ago. It was mimeographed, had 64 pages and contained works from only two magazines.

With the passage of time our objectives became known to a wider circle and today we are priviledged to be able to include road tests and other articles from over thirty journals. These stories pen an international picture of a particular model as they draw together views from as far apart as Sydney and London as well as Los Angeles and Detroit.

The reason that there are now nearly 250 titles in our series is that as vehicles progress from hand to hand their new owners thirst for information about their out of production cars. They want to know how it performed when it left the factory, how it was received by the press of the day, how it evolved and did it triumph on the track.

Fortunately for these enthusiasts all the leading magazine publishers understand their needs and have for many years generously allowed us to pick up their copyright technical articles and reprint them in small numbers in this format.

We are sure that Nova devotees will wish to join with us in thanking the management of Canada Track and Traffic, Car and Driver, Car Craft, Car Life, Cars, Hot Rod, Motorcade, Motor Trend, Road Test and Wheels for their contribution to the hobby and for their continued support.

R.M. Clarke

Driving the CHEVY II

A first turn behind the wheel of the new-size Chevrolet reveals the results of some interesting engineering

by Jerry Titus

EACH SUMMER, long before most folks begin thinking about next year's automobiles, Detroit cracks its factory gates open to journalists for a brief advance peek at the coming line. Often, in addition to the look, there is a chance to slip behind the wheel of a prototype or a pilot production model and drive it for a limited time on a proving ground. That is exactly what we were able to do with the Chevrolet II this year. There wasn't time for a full-scale, get-acquainted road test (which will come later), but it was possible to take an objective ride in Chevy's new line, size it up, and do a little early-season prognosticating on its success potential.

The Chevy II comes in three series: 100, 300 and Nova. We had the top-of-the-line Nova convertible. Power is from a 120-hp, 194-cubic-inch, in-line Six, the only engine available for the Nova. This same unit is a $55 option for the 100 and 300; standard is a 153-inch, 90-hp Four.

The Chevy II is on one hand a most straightforward car — simple, honest and conventional — rather obviously designed to compete directly with the Falcon, whose dimensions it closely parallels. On the other hand, it introduces the single-leaf rear spring, one of the most advanced suspension ideas to be seen in years. It is this unusual longitudinal spring, two per car, that gives the II much of its handling and riding personality. Very briefly, it varies in width and thickness throughout its five-foot length, is sandwiched in rubber at the axle (no holes in the spring) and is rubber bushed at the ends.

How it actually works seems almost contradictory. There is a great deal of body roll, but the car does not feel unstable. The ride is soft and pleasing, but not seasick-soft with the constant pitching and rolling that some cars have. In using a lightweight spring, Chevrolet engineers have gained a more favorable ratio between sprung and unsprung weight, which definitely helps the ride.

From the outside one gets the impression that the II is a small car, nicely proportioned, but definitely small. Once inside, that impression is knocked right out the door. Clever design work has kept usable interior space to a maximum, considering the car's overall dimensions.

Our Nova was easy to drive, easy covering two essential functions — steering and braking. Non-power steering is on the slow side (4¾ turns, lock-to-lock), but it was extremely light, which made the car easy to park and a one-finger exercise to steer while moving. Power-assisted brakes required light pressure and did their job without locking during hard stops.

CLEAN, SIMPLE, STRAIGHTFORWARD STYLING GIVES THE NOVA CONVERTIBLE PLENTY OF EYE-APPEALING CLASS FOR A COMPACT-SIZE CAR.

When power is rated at 120 horses and the curb weight is 2850, the result is moderate performance. With the Powerglide transmission (three-speed manual is standard) and a 3.36-to-1 rear axle, we were able to arrive at a 0 to 30 time of just over five seconds, about 10 seconds to 45, and a shade under 16 seconds to 60. Although top speed is 98 mph, the car seems at its best below 75, where it did not feel as though it were working hard. Fuel consumption appeared as though it would average out about 23 mpg.

Although the four-cylinder engine doesn't go with the Nova line, it is important in the overall II picture. It is a sturdy engine, designed for wide-open running without injury. To place it in performance perspective, acceleration to 60 mph should take about 20 seconds.

Factory list prices range from $1827 for a two-door sedan to $2282 for a Nova wagon. We think we are moving out on a sound limb by predicting that the II will push rapidly into the top level of the compact car sales picture. /MT

All-new, cast-iron in-line Six pulls 120 horses from 194 cubic inches. Standard in the Nova line, it is an extra-cost option on the 100 and 300 series.

CT&T ROAD TEST

Chevy II *An Honourable Compromise*

G.M.'S "COMPROMISE SIZED" Chevy II, like early-1950's Chrysler Corporation cars, is small on the outside, big on the inside. Intended to fight Ford's fast-selling Falcon in the utility sedan market, the Chevy offers big-car room coupled with compact-style economy and maneuverability.

Customers can choose between a 120-hp, 6-cylinder or 90-hp, 4-cylinder powerplant, Chevy's first in-line 4 since 1928. Automatic transmission is available to replace the standard 3-speed manual shift, along with such other extras as power steering and brakes. Our test car was equipped with the 6-cylinder engine and automatic transmission.

Shortly after introducing the Chevy II, G.M. of Canada announced the "Acadian". Despite the "totally new" and "all Canadian claims", the Acadian is simply a Chevy II with petty trim alterations and the usual maple leaf insignia traditionally bestowed on "Canadian" cars.

TECHNICAL

The Chevy II's humble aspirations are best realized by straightforward, proven engineering and the car carries few breathtaking innovations.

Front suspension is by coils, while at the rear a new wrinkle is tried with "tapered plate springs".

The spring in question is a single leaf type, five-foot shot-peened steel bar varying in thickness and width for uniform stress distribution and mounted in heavy rubber insulators. Purpose is to give rear axle cushioning which cuts weight and assembly time with no loss in ride quality.

Integral body-frame construction is another interesting feature, and a unitized front end assembly with bolt-on fenders should lighten repair work and bills.

STYLING

Thoroughly American in concept, the Chevy II still bears more than faint resemblance to European sedans with sharp-edged looks, squarish corners and an almost aggressive simplicity. The abruptly squared-off tail carries two of the smallest brake lights of any U. S. car since the early 50's.

Overall the car comes off as an honest, commendably neat job if not an inspiring beautiful one. Workmanship was unexceptional but satisfactory on our test car.

INTERIOR

Bench-type seats front and back are firm, upright and upholstered in subdued textured fabric with vinyl trim. The roominess of this interior made us wonder why cars need to be any bigger. We noted a pleasing lack of tinny glitter which helped create a feeling of restrained good taste — a welcome departure from current practice. Materials seemed to be of generally durable quality except for the glove compartment, the interior of which is nothing more or less than a stapled cardboard box.

Because styling hasn't overwhelmed function completely, the Chevy II's instrument panel is one of the better designs seen of recent years on an American car. Instruments include a speedometer (a welcome revival of the old-fashioned circular variety), fuel gauge, and lights for temperature, oil and generator — all mounted in a panel where they can be seen unobstructed through the steering wheel.

Visibility is first rate and there isn't a radically curved piece of glass anywhere. With its short hood, high seating position and near-vertical steering wheel placement, the car seems built for easy and alert driving.

DRIVING

Our test car started up instantly in frosty weather with a single turn of the key. Powerglide automatic transmission as supplied on our test car worked smoothly except for a noticeable jerk at about 30 mph as it shifted up under harsh acceleration. In night driving, the shift quadrant isn't illuminated and some difficulty was encountered in selecting the proper setting. Also, it's a dangerously easy matter to swing the quadrant from Low right over to Neutral, missing the Drive position entirely — not a recommended practice when the gas pedal is floored.

Pep aplenty is forthcoming from the 194-cu. in. 6-cylinder Chevy II engine. Not enough to produce screeching wheelspin but more than ample for fast traffic getaways and passing reserve on the highway. Noise is limited to a rather pleasant hushed rasp, although at cruising speeds a fair amount of wind-buffeting produced various whistles and hums.

Steering is inclined to be somewhat heavy in tight low-speed maneuvers but is not noticeably difficult once above 30 mph. Power steering would be pleasant but we didn't really feel it necessary at any time. Most standard-size American cars are sheer boredom for the sports-minded driver, but the Chevy II is a long way from being drudgery. Handling was a real surprise; with a minimum of squeal and lean, the Chevy II corners neatly and adroitly and has the manners of a much smaller, lighter automobile. Far more correction is needed than on smaller cars, but this car can be briskly thrown about to a degree that curving roads are welcomed. Really fast corners reveal gradual understeer that gives plenty of warning before the tail begins wagging, but with the rather loose steering characteristic an incipient spin would be a handful to

Neat hindquarters add to European look

Grille shows Chevy II is a Chevy too.

Canada Track & Traffic/December, 1961

correct in time. There isn't the sheer guttiness needed to power out of such situations, so though the Chevy is definitely fun to drive we wouldn't recommend it wholeheartedly for race-track type jollies.

Brakes, without power assist, efficiently served to haul the car down to straight and rapid stops.

Not harsh by any means, the Chevy II's ride is a good deal firmer than that of its larger domestic contemporaries. The suspension, however, and not the steering wheel absorbs the blows. Choppy surfaces produce some jounce but no rattles.

ECONOMY

Since its *raison d'etre* is purely and simply economy, the Chevy II was expected to perform well in the mileage department. Some strenuous driving notwithstanding, we recorded 22 mpg in combined city and open-road driving.

STORAGE

Its boxy lines have given the Chevy a wide-mouthed, rectangular trunk that should devour all sane luggage requirements on long trips.

HEATER AND VENTILATION

Accompanied by an inexcusable amount of fan noise, the heater works up warm air quickly and via outlets under the instrument panel gets it to the occupants rapidly. Fresh air vents near the firewall let in cold air when needed.

LAST WORD

There's nothing pretentious about the Chevy II and in that light it appears to be a sound buy. Its fresh looks and traffic agility are strong points, as are its roominess and satisfactory if not sensational operating economy. A few signs of flimsy construction mar our unqualified endorsement, but discounting the flaws in our particular test car the Chevy II is a sturdy, straightforward utility car which should appeal to the economy-conscious buyer who still wants to enjoy driving.

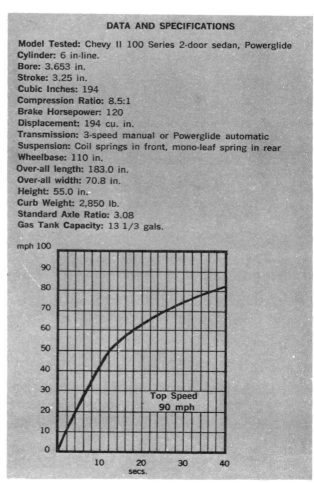

DATA AND SPECIFICATIONS

Model Tested: Chevy II 100 Series 2-door sedan, Powerglide
Cylinder: 6 in-line.
Bore: 3.653 in.
Stroke: 3.25 in.
Cubic Inches: 194
Compression Ratio: 8.5:1
Brake Horsepower: 120
Displacement: 194 cu. in.
Transmission: 3-speed manual or Powerglide automatic
Suspension: Coil springs in front, mono-leaf spring in rear
Wheelbase: 110 in.
Over-all length: 183.0 in.
Over-all width: 70.8 in.
Height: 55.0 in.
Curb Weight: 2,850 lb.
Standard Axle Ratio: 3.08
Gas Tank Capacity: 13 1/3 gals.

Top Speed 90 mph

FOUR CYLINDER engine is first offered by Chevrolet since 1928. Listed as standard for the Chevy II, it puts out 90 bhp from its 153 cu. in. displacement. Shroud ducts air to radiator.

CHEVY II

Car Life's Award car undergoes scrutiny in all four forms: 4- and 6-cyl. with manual and automatic transmissions. Each has its particular set of advantages. Each emphasizes one point—it's possible to build a good small car in the U.S.

MODERN SIX is optional in the II, standard in the Nova. It boasts rugged, 7-main-bearing construction and its 194 cu. in. delivers 120 bhp. Both four and six have hydraulic valve gear.

WITH THE *Car Life* engineering award going to the Chevy II, we felt that a road test report on all four basic models would be appropriate. The engineering may be excellent, but does anything suffer in the execution? What are the cars like on the road? How do they perform, is the 4-cyl. version adequate, how do they ride, are they cheaper to operate, and what are the drawbacks, if any?

Since *Car Life* already has a reputation for never liking any car 100%, let us say at the outset that the Chevy II is no exception to our rule. Our first road test of a Chevy II appeared in CL for November, 1961, and was a 6-cyl. model with automatic transmission. The question we wanted to answer was: How do both the 4 and 6 perform, and with either the standard shift or optional Powerglide automatic transmission?

In a very quick performance comparison the figures

Car Life Road Test

look like this:

0-60 MPH

Chevy II, 4-cyl. automatic 20.0 sec.

same, manual 17.0 sec.

Chevy II, 6-cyl. automatic 14.5 sec.

same, manual 13.0 sec.

Those figures pretty well tell the performance story. Everything else follows the same pattern: top speed, hill-climbing ability, high gear acceleration, quarter-mile elapsed time, etc. For a more complete comparison see the accompanying box score tabulation.

CHEVY II

The 4-cyl. Chevy II

When Chevrolet introduced the Corvair two years ago it anticipated that the big demand would be for the lower-priced, plain-Jane series 500 model with synchromesh transmission. This proved to be a tactical error which slowed early sales. Customers waited to get delivery on more de luxe Corvair 700s with automatic transmissions. This time, with the Chevy II, the opposite policy was in effect. We had a very difficult time in arranging for a test of the 4-cyl., manual-shift model. There weren't any.

The manual-shift 4 really is a serious threat to the imported car market; it is sensible, practical, economical, basic transportation with the added advantage of full-size interiors, a better ride and a nationwide network of dealers. There are a few drawbacks, however, and we may as well point them out—you'll know what we mean when you try one yourself.

The 4-cyl. engine is somewhat more noticeable in the manual-shift model, as compared with a Powerglide-equipped sedan. A part of this problem can be attributed to the time-tried but odd synchromesh design. Used by Chevrolet since 1937 and beefed up in 1955, it has good ratios but always was noisier than most, in 1st and 2nd gears. The engine feels just a little rough as you start to move off and, though smooth enough up to about 70 mph, it somehow is just a little noisy. A cruising speed of around 70 mph (an actual 65, both 4s had very optimistic speedometers) is very pleasant but above that speed the engine begins to feel too busy, even though we know that it really will take an all-day thrashing at 80 mph.

The 4 will idle down to as low as 20 mph in high gear but will not accelerate smoothly from this speed, and 25 mph is the minimum comfortable speed. Anything below that requires 2nd gear, which, by the way, is good for over 60 mph if you want to push hard. First gear is well chosen for best possible acceleration from a standstill and the Chevy II-4 will burn a little rubber if you have no one in the rear seat. However, in comparison to the imported competition mentioned earlier, it is unfortunate that the 4-cyl. Chevy II is not available with the superb Corvette all-synchromesh 4-speed transmission.

In brief, if you want the extra economy of only 4 cyl. the manual transmission is the better buy for chiefly highway travel, whereas the automatic transmission has definite advantages in the city.

4-cyl. Chevy II with Powerglide

Most of our objections to the 4-cyl. model are alleviated by the optional Powerglide transmission. However, the P-G defeats its own purpose in being expensive ($168 extra). An economy car (a bad word around Detroit) must be low in first cost, easy on gas and cheap to maintain. The Powerglide is not only expensive, it also increases the fuel consumption rate. For cross-country driving the increase is not important; no more than a drop of 1-2 mpg. But around town the mpg drops to well below 20. Because we got the manual-shift car so late, we can't give any precise comparisons although we recorded 21.8 mpg for 300 miles of assorted driving with the Powerglide 4. About 100 miles of this was in the city, and the rest was expressway cruising at 65-70 mph. The Powerglide 4 gives a low of 19 mpg, with a high of about 23-24 mpg. The manual shift 4, with its 3.08 axle ratio (changed in subsequent production to 3.55 for all models) should never fall below 20 mpg under the worst conditions and should get an easy 25 mpg or more on the highway.

Most of our objections to 4-cyl. roughness disappear with the automatic transmission. There is no engine shake, no problem with low speed bucking and jerking. On acceleration the initial take-off is slower, despite the much

| TABULATION OF PERFORMANCE |||||
| Performance Summary |||||
No. cylinders	4	4	6	6
Transmission	auto	stick	auto	stick
Test weight	2960	2940	3150	3130
Axle ratio	3.36	3.08	3.08	3.08
Acceleration				
0-30	6.0	4.9	5.5	3.8
0-60	20.0	17.0	14.5	13.0
0-80	45.0	41.0	28.5	27.0
ss ¼	21.7	20.2	19.5	19.0
Top speed	82	84	92	94
Tapley pull	210	205	260	245
Mpg, city	19	21	17	18
Mpg @ 50	24	25	22	23

more favorable over-all ratio of 14:1. The low-to-high upshift takes place at 46-47 mph under wide open throttle (4200 rpm) and the 0-60 time is 3 sec. longer than for the manual shift. It is possible to hold low range up to nearly 60 mph but this procedure doesn't improve the zero to a *corrected* 60 mph time because of a rather slow, cushioned upshift (see later remarks on the 6-cyl. models where the situation is different).

The ride of the 4-cyl. cars is very good in comparison with similarly priced imports, but the front end springing in particular feels noticeably harsher than the 6-cyl. models, which have more weight in front (and different springs) but the same ride rate (120 lb./in.). Being accustomed to a Corvair, we disliked the feel of the Chevy II in a corner—you sit up so much higher and there is much more body roll. But the car will corner satisfactorily once you get used to the roll. The moderate understeer is very comforting on high-speed bends. The Chevy II is also much less affected by cross winds than its rear-engined brother.

In all, we drove three 4-cyl. cars and two of them had power steering. This is a ridiculous option ($75 extra) for the 4-cyl. car and the steering ratio is unchanged at 4.7 turns lock to lock. Even the 100-lb.-heavier 6-cyl. cars are light and easy to park with manual steering—with the added benefit of better control and general feel-of-the-car on mountain roads.

Four of the six cars we drove had power brakes. The only difference we could detect was that the non-boosted brakes seemed to take more pedal-pressure when the car was brand new but after 500 miles the linings seem to seat in, and on cars with several thousand miles it seemed to us that the difference in pedal pressure was negligible, or at least hardly worth the money ($43). In general, the brakes are adequate though a trifle erratic until properly worn in. The car will pass our 80-mph crash stop test with almost no fade, but not a second such test, performed immediately afterward.

Sintered metallic brake linings are optional and might be advisable for fast, hard driving.

The Manual-shift 6

Our manual-shift 6 was the super-de luxe Nova 400 sport coupe, which has oversize tires (6.50-13) as standard equipment because it is nearly 100 lb. heavier than a 6-cyl., 4-door sedan. Thus a standard sedan would perform slightly better than recorded here for two reasons: 3.5% less weight and 4.3% more engine revolutions per mile.

Although one Chevrolet owner on our staff refers to this particular model as "suddenly it's 1940," we liked it the best of all. Surprisingly, it does compare in some ways to the pre-war standard size Chevrolet—for example:

	1940	1962
Wheelbase	113.0	110.0
Curb weight	3040	2850
Tire size	6.00-16	6.50-13
Displacement	216.5	194.0
Bhp	85	120
Axle ratio	4.11	3.08
Engine revs/mile	3020	2620
0-60 time	18.8	13.0
Top speed	80/82	92/94

But the big step forward, exemplified by the 6-cyl. Chevy II, is in performance, economy and riding quality. The 22-year-old car wouldn't do over 55 mph in 2nd gear, while the new one does 65 easily and, if pressed to the absolute limit, will touch 70 mph. The old car would sometimes get 17 mpg on a trip, and the new one gets 18-20 mpg, consistently. And, of course, the riding qualities of the Chevy II-6 are much improved and the car is much easier to handle.

As with the 4, the manual transmission is a little noisy. This is not as noticeable under hard acceleration through the gears as it is when cruising along in slow traffic, where the backlash shows up and there is a different gear noise on the over-run from that when pulling. The minimum speed in high gear is 20 mph and the engine pulls away from this speed without strain. Low or 1st gear is almost too low; during our acceleration checks it was necessary to experiment in order to get the best possible starts without too much rear wheel spin. Second gear starts are a little hard on the clutch, but if the car is barely rolling this gear is quite adequate.

Despite a very slow-turning engine, or maybe because of it, the car is at its best on the open highway. The engine is quiet and there are no vibration periods. A speed of 80 mph is pleasant and high-gear acceleration up to this speed is brisk, though a little slow above that point. The speedometer proved to be very accurate, in fact slightly slow in the upper ranges. Over 90 mph is possible, and we once

FEBRUARY 1962

CHEVY II

saw 95 mph with a brisk tail wind. In this connection, we suspect that Chevrolet does not change speedometer gears with tire size; if correct, this would explain why one of our earlier test cars had a 3% error (fast) while this one was slow. (Tire companies give the 6.00-13 tires' revolutions as 890 per mile, while the 6.50-13 size is rated at 850 revs/mile.) Strangely enough, the smaller tires definitely make the 6-cyl. car faster: about 94 mph under favorable circumstances. However, those interested in performance should order the 6-cyl. Chevy II with the 6.50 tires and the optional 3.36 axle ratio; this would give a theoretical 94 mph at 4450 rpm.

The 6-cyl. Powerglide

The 6-cyl. Chevy II with Powerglide transmission will undoubtedly be the biggest seller, despite the extra cost of $223 ($55 for the engine, $168 for Powerglide). It is a delightful car to drive and most owners can expect a tank mileage of close to an honest 20 mpg, in normal everyday driving.

Unlike the 4-cyl. Powerglide model, the 6-cyl. car can be held in low range to considerable advantage, if better performance is desired. In repeated tests we averaged a 0-60 time of 13.5 sec. by merely using low range all the way. This requires 4750 rpm with 6.50-13 tires. Earlier tests, with a lighter 4-door sedan and 6.00-13 tires re-

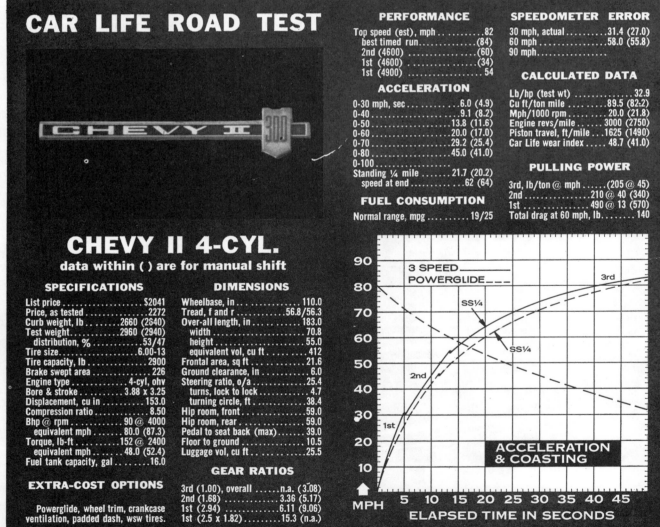

CAR LIFE ROAD TEST

CHEVY II 4-CYL.
data within ( ) are for manual shift

SPECIFICATIONS
List price	$2041
Price, as tested	2272
Curb weight, lb	2660 (2640)
Test weight	2960 (2940)
distribution, %	53/47
Tire size	6.00-13
Tire capacity, lb	2900
Brake swept area	226
Engine type	4-cyl, ohv
Bore & stroke	3.88 x 3.25
Displacement, cu in	153.0
Compression ratio	8.50
Bhp @ rpm	90 @ 4000
equivalent mph	80.0 (87.3)
Torque, lb-ft	152 @ 2400
equivalent mph	48.0 (52.4)
Fuel tank capacity, gal	16.0

EXTRA-COST OPTIONS
Powerglide, wheel trim, crankcase ventilation, padded dash, wsw tires.

DIMENSIONS
Wheelbase, in	110.0
Tread, f and r	56.8/56.3
Over-all length, in	183.0
width	70.8
height	55.0
equivalent vol, cu ft	412
Frontal area, sq ft	21.6
Ground clearance, in	6.0
Steering ratio, o/a	25.4
turns, lock to lock	4.7
turning circle, ft	38.4
Hip room, front	59.0
Hip room, rear	59.0
Pedal to seat back (max)	39.0
Floor to ground	10.5
Luggage vol, cu ft	25.5

GEAR RATIOS
3rd (1.00), overall	n.a. (3.08)
2nd (1.68)	3.36 (5.17)
1st (2.94)	6.11 (9.06)
1st (2.5 x 1.82)	15.3 (n.a.)

PERFORMANCE
Top speed (est), mph	82
best timed run	(84)
2nd (4600)	(60)
1st (4600)	(34)
1st (4900)	54

ACCELERATION
0-30 mph, sec	6.0 (4.9)
0-40	9.1 (8.2)
0-50	13.8 (11.6)
0-60	20.0 (17.0)
0-70	29.2 (25.4)
0-80	45.0 (41.0)
0-100	
Standing ¼ mile	21.7 (20.2)
speed at end	62 (64)

FUEL CONSUMPTION
Normal range, mpg	19/25

SPEEDOMETER ERROR
30 mph, actual	31.4 (27.0)
60 mph	58.0 (55.8)
90 mph	

CALCULATED DATA
Lb/hp (test wt)	32.9
Cu ft/ton mile	89.5 (82.2)
Mph/1000 rpm	20.0 (21.8)
Engine revs/mile	3000 (2750)
Piston travel, ft/mile	1625 (1490)
Car Life wear index	48.7 (41.0)

PULLING POWER
3rd, lb/ton @ mph	(205 @ 45)
2nd	210 @ 40 (340)
1st	490 @ 13 (570)
Total drag at 60 mph, lb	140

ACCELERATION & COASTING

corded 13.0 sec. (5000 rpm). So we can say that a Powerglide 6, properly driven, can very nearly equal the acceleration performance of the manual-shift version.

But acceleration to 60 mph isn't everything, and a time of 14.5 sec. with automatic upshift at 55 mph isn't slow by any standard. The Powerglide model also gets extra marks because its high gear pulling power, as demonstrated by the Tapley meter, is definitely better than with manual transmission. This, of course, means that the P-G 6 will climb a steeper hill in high gear without shifting down.

In this connection, the Tapley meter is an accurate measure of genuine pulling power or engine torque. In all four cars we were surprised at the high reading which came up at low speeds and continued to hold steady up to around 55 mph (where wind resistance takes over). This proves that these engines give exceptional torque, low-down, and the torque curve stays on—long and flat.

But, reverting to the Powerglide, this unit is technically a 2-speed plus converter transmission. What is often overlooked is the true effect of the converter. This portion of the transmission gives another gear ratio and in the Chevy II we find a new design which gives more torque multiplication than before. The over-all break-away ratio is stall torque ratio x transmission ratio x axle ratio, in this case 2.50 x 1.82 x 3.08, or 14.0 overall (16.4 with the new ratio) and, despite some slip and thermal losses, the effect is somewhat better than with the 3-speed manual transmission with its over-all break-away ratio of 9.06. The full throttle upshift occurs at 4350 rpm, practically at the peak power point. Then the engine drops back to 2400 rpm and, even with some converter slip, this speed is right on the torque peak. It's all pretty efficient, even though acceleration from, say, 55-70 mph could be improved by a 4-speed manual transmission.

Summary

The 4-cyl. cars offer adequate though not brilliant performance and the manual-shift version is much better in both performance and economy if you don't mind shifting gears.

As we said earlier, the 6-cyl. models with manual shift perform brilliantly and are the most fun to drive. We recommend the optional 6.50 tires (which come on wider rims), because the 6.00 tires are marginal for the load imposed. An axle ratio of 3.36 can be ordered to offset the larger tires if performance is your objective.

The Powerglide 6-cyl. will suit the greatest number of people, though fully equipped it will cost you more than a plain Biscayne 6. But the Chevy II has a tremendous advantage in its handier, easy-to-drive package, plus obvious economies in gas, oil, tires, etc. If you want a big, heavy car, be prepared to pay for it—but the Chevy II will be less expensive in the long run. "Weight is the enemy" is an old axiom—and it still stands. ■

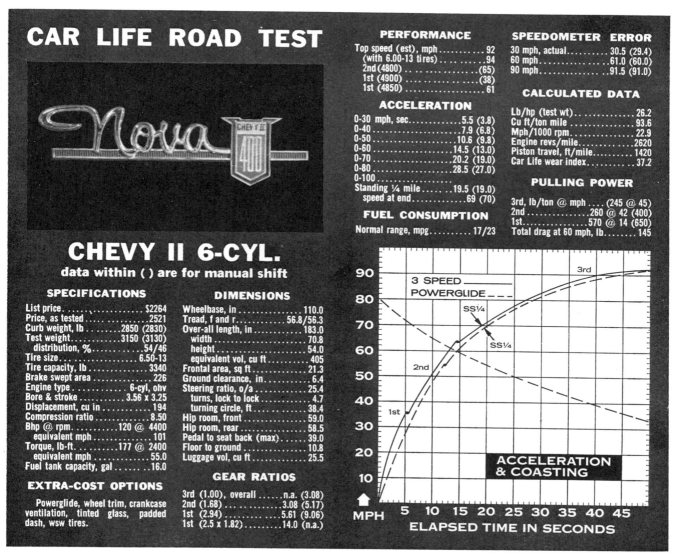

CAR LIFE ROAD TEST

CHEVY II 6-CYL.
data within ( ) are for manual shift

SPECIFICATIONS
List price	$2264
Price, as tested	2521
Curb weight, lb	2850 (2830)
Test weight	3150 (3130)
distribution, %	54/46
Tire size	6.50-13
Tire capacity, lb	3340
Brake swept area	226
Engine type	6-cyl, ohv
Bore & stroke	3.56 x 3.25
Displacement, cu in	194
Compression ratio	8.50
Bhp @ rpm	120 @ 4400
equivalent mph	101
Torque, lb-ft	177 @ 2400
equivalent mph	55.0
Fuel tank capacity, gal	16.0

EXTRA-COST OPTIONS
Powerglide, wheel trim, crankcase ventilation, tinted glass, padded dash, wsw tires.

DIMENSIONS
Wheelbase, in	110.0
Tread, f and r	56.8/56.3
Over-all length, in	183.0
width	70.8
height	54.0
equivalent vol, cu ft	405
Frontal area, sq ft	21.3
Ground clearance, in	6.4
Steering ratio, o/a	25.4
turns, lock to lock	4.7
turning circle, ft	38.4
Hip room, front	59.0
Hip room, rear	58.5
Pedal to seat back (max)	39.0
Floor to ground	10.8
Luggage vol, cu ft	25.5

GEAR RATIOS
3rd (1.00), overall	n.a. (3.08)
2nd (1.68)	3.08 (5.17)
1st (2.94)	5.61 (9.06)
1st (2.5 x 1.82)	14.0 (n.a.)

PERFORMANCE
Top speed (est), mph	92
(with 6.00-13 tires)	94
2nd (4800)	(65)
1st (4900)	(38)
1st (4850)	61

ACCELERATION
0-30 mph, sec	5.5 (3.8)
0-40	7.9 (6.8)
0-50	10.6 (9.8)
0-60	14.5 (13.0)
0-70	20.2 (19.0)
0-80	28.5 (27.0)
0-100	
Standing ¼ mile	19.5 (19.0)
speed at end	.69 (70)

FUEL CONSUMPTION
Normal range, mpg	17/23

SPEEDOMETER ERROR
30 mph, actual	30.5 (29.4)
60 mph	61.0 (60.0)
90 mph	91.5 (91.0)

CALCULATED DATA
Lb/hp (test wt)	26.2
Cu ft/ton mile	93.6
Mph/1000 rpm	22.9
Engine revs/mile	2620
Piston travel, ft/mile	1420
Car Life wear index	37.2

PULLING POWER
3rd, lb/ton @ mph	(245 @ 45)
2nd	260 @ 42 (400)
1st	570 @ 14 (650)
Total drag at 60 mph, lb	145

FEBRUARY 1962

Although it is being sold here in limited quantities; the Chevy II has more than usual significance for Australian motorists as it is the basis for the new Holden.

CHEVROLET'S COMPACT CHEVY II

By PETER HALL

THERE are plenty of rumors around already allegedly giving the complete lie to what the next Holden will be like and when it will come. I prefer to wait till GM-H announces the final details — in the past they have been the only ones who have been right.

But it does seem obvious that fairly soon, the Holden it going to get at least a brand new body. And the most intelligent birds around the industry are tipping it will be like a new American GM car that was released late in 1961 — the Chevy II.

Interestingly enough, the Chevy II was developed by GM to undermine Ford's Falcon — easily the best-selling compact in the United States and a much more popular car than GM's original entrant in the compact field — the rear-engined Corvair.

So, my ear cocked a little higher than usual upon hearing these things, when the news came my way that a handful of Chevy II's were being imported to Australia — by Preston Motors in Melbourne and Stacks in Sydney — as a low-volume special for their well-cashed Chevrolet clientele.

Preston Motors were good enough to register the first one they got, run it in as a demonstrator and hand it over to me for what proved to be a fascinating test. Before I go into a full description of the test, it might be well to state a few facts about the Chevy situation in America. There, the basic Chevy II is a four cylinder job that develops 90 brake horsepower, has a three speed column shift transmission, few extras and sells for slightly less than the Falcon— around the £900 (Australian) mark.

Various options are offered, including a six-cylinder engine, the familiar 2-speed Powerglide automatic transmission that has been fitted to the Australian assembled big Chevs since 1959 and fancier trims and equipment. There are no less than nine models of the Chevy II on sale in America, including station wagons, a hardtop coupe and a convertible.

Knowing well that the hefty import duties that apply to all-imported American cars would put the Chevy II way out of the Falcon bracket here, the Australian importers went for the most luxurious

Big, curved rear window and absence of fins shows car's relation to bigger Impala models.
WHEELS, July, 1962

Clean frontal treatment, high ground clearance are features of the Chevy II, Chevrolet's second compact car.

version of the Chevy, with quite a few of the optional fittings, and decided to sell it as a slightly cheaper Chevrolet. Hence the whopping £2150 price tag.

The test car was fitted with the six cylinder engine, as are all the Chevy II's being brought out here.

A 3.18 litre unit very similar to the six-cylinder Chevs that were so well known to Australians for three decades before the switch to V8's, the engine develops a healthy 120 bhp.

Transmission is the 2-speed automatic, and such trimmings as a tinted windscreen, two-speed electric windscreen wipers, wheel discs, cigarette lighter and windscreen washers are thrown in as part of the price.

But — believe it or not — there is no heater. And the floors are covered in a dreadful rubber substance moulded into thousands of ugly little mounds that, I understand, are supposed to give the impression that the stuff is "imitation carpet". The plain metal floor would have looked better.

From the outside, the Chevy II is a good looking package without going to extremes. Modern without being garish and practical but not dull, the looks of this car would make any Holden owner beam appreciatively.

In detail, there are many similarities to the Falcon, but the car, thanks to clever design, does not look at all like its main American rival in the overall picture. It is squarer and the shape is not dominated by enormous areas of glass back and front.

Nevertheless, both the windscreen and back window are large, really large, and give the driver excellent vision.

The interior of the test car was colorful in the American tradition, with lashings of red paint, red leather strips and red patterned plastic inserts in the seats. Even the steering wheel was red. And, of course, the imitation carpet, too.

The doors open wide, the seats are well shaped and the exceptionally well placed and smallish steering wheel allows the driver to slip in without effort. There is ample room for six people, and even the transmission tunnel is a moderate height.

The dashboard rolls back and forth like an indecisive pinball, but the matt finish on top of it is a welcome idea that prevents reflections in the windscreen. The instruments — a speedo, fuel gauge and the usual GM warning lights, are right in front of the driver and are easily read through the two-spoke steering wheel. Control buttons are arrayed along the lower edge of the dash on either side of the steering column, within easy reach of the driver.

The driving position itself is one of the best I have come across in an American car. The driver is well supported by the seat, is high enough to get a good view of the road about him and is far enough back from the wheel to allow him to have complete control over it. The pedals — a broad brake pad and organ-

WHEELS FULL ROAD TEST

Chevy IIs being imported here have all luxury options — tinted windscreen, two-speed wipers, cigarette lighter.

type accelerator — are well placed. But the dipper switch, so often placed in awkward positions, was on the flat of the floor against the transmission tunnel —in about the most uncomfortable and least accessible place it could be.

Though it is lower than the Holden, the Chevy II beats the Australian job for room and comfort for those inside it. Even in the back seat with the front one pushed right back, there is adequate leg room, and one's head is a comfortable distance from the roof.

Suspension proved one of the Chevy II's most interesting features. The front springing seems a close copy of the British Ford system first used on the Zephyr and Consul many years ago and is now very familiar to Australians as the front springing on the Falcon.

It is independent with the coil springs set high under the mudguards, above the upper A arms of unequal length. An anti-roll bar is fitted and the whole front suspension is designed with inbuilt antidip forces.

The rear suspension is quite unique on modern cars in that the semi-elliptic springs in the otherwise quite conventional solid axle set-up each have only one leaf.

The suspension certainly showed up in very good light on the test run. Soft enough for comfort around the city at moderate speeds, it is sufficiently firm to hold down the Chevy at speed on the open road.

Furthermore, it is possible to belt the car across really broken outback roads at highway speeds without any danger of the driver losing his grip on things and without a trace of the sideways crab-action that characterises a lot of solid back axle cars under similar conditions.

The single leaf back springs seem free of vices. They are silent, held the car on a level keel and look strong enough to stand up to a lot of rough usage. Indeed, the only time I was aware that they could have been acting any differently from conventional leaf springs was when I hurled the Chevy through a tight S bend right on the limit. Then, as I flicked the car over to scurry round the second half of the S, the back of the car seemed to spring across on to its opposite keel, as if the back springs had wound themselves around like a spring and unwound in a hurry at the first sign from the driver that that was what they were expected to do. It was an unusual sensation first time it happened, but it helped rather than marred the car's excellent cornering ability.

The steering is very American — soft, very low geared, with strong self-centring action and a lot too

WHEELS, July, 1962

many (4.7) turns from lock to lock for the size and weight of the car. But, it is very light and parking is a cinch.

With the 120 bhp engine fitted, the Chevy II turned in a most impressive performance. Acceleration was hard and clean, despite the handicap of a two-speed automatic transmission, and the car had no trouble getting to and maintaining 90 mph. As it was becoming a little light in the front and a little too sensitive to a mild cross wind a few mph before then, I doubt if a higher top would be needed by anyone not intending to race it.

Brakes are a lot better than is customary for big American cars. They took many hard stops without fading and when fade did set in, it was not total and by no means permanent. Pedal pressure is light and the braking action pleasantly progressive.

The quality of finish and the care with which the body was put together is not up to Australian standards, but it was not too bad compared with some late model American examples I have seen.

Paintwork — a handsome irridescent grey with a white top — was smooth but looked a little "thin", and you did not have to look far to come across ugly weld spots. Insulation from road noises is effective and there is wind whistle only when the quarter vent is opened at speed. But the insulation from engine noise is not very effective at all.

One of the nicest "little" things about the Chevy II is the windscreen wiping system. Electrically operated, the blades run in parallel arcs with the choice of two speeds. The blades swept almost all the moderately curved windscreen.

In the form in which I tested the Chevy, with its fairly big six cylinder engine, the car was about 1¼ cwt heavier than the current EK Holden. But, with the 4-cylinder engine fitted in the standard Chevy II in USA, the weight difference comes down to about 40 lb. There is not much difference in weight between the Chevy's big four and Holden's light six. Aerodynamically, the Chevy is clearly ahead of the Holden.

And, it is approximately the same size. To be exact, the Chevy II is two inches longer, two inches wider and 1.2 inches lower, but it has a 5 in longer wheelbase.

Apart from this purely Australian interest in the Chevy II, it is a very interesting car in its own right.

It has power, performance, reasonable economy and more than enough space and good looks for the average man. It is a particularly good example of a trend that makes one begin to believe that American car makers are regaining the sanity and good sense that once put them so far ahead of everyone else. #

Test car had 120 bhp, 3.18-litre engine similar in design to ones used by Chevrolet before the V8s were adopted.

wheels ROAD TEST

TECHNICAL DETAILS OF THE CHEVY II

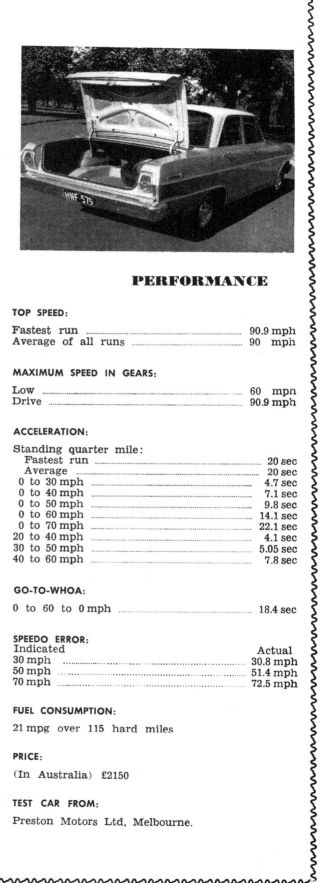

SPECIFICATIONS

ENGINE:

Cylinders	six, in line
Bore and stroke	3.563 x 3.25 in
Cubic capacity	3179 cc (194 cu in)
Compression ratio	8.5 to 1
Valves	overhead, pushrod, hydraulic tappets
Carburettor	single downdraught
Power at rpm	120 at 4400

TRANSMISSION:

Type .... Powerglide, 2-speed automatic

SUSPENSION:

Front	independent, coil springs set above upper wishbones
Rear	solid axle, single leaf semi-elliptic springs
Shockers	telescopic

STEERING:

Type	recirculating ball
Turns, lock to lock	4.7

BRAKES:

Type .... hydraulic, drums

DIMENSIONS:

Wheelbase	9 ft 2 in
Track, front	4 ft 8.8 in
Track, rear	4 ft 8.3 in
Length	15 ft 3 in
Width	5 ft 10.8 in
Height	4 ft 7 in

TYRES:

Size .... 6.40 x 13

WEIGHT:

Kerb .... 23¾ cwt

PERFORMANCE

TOP SPEED:

Fastest run	90.9 mph
Average of all runs	90 mph

MAXIMUM SPEED IN GEARS:

Low	60 mph
Drive	90.9 mph

ACCELERATION:

Standing quarter mile:

Fastest run	20 sec
Average	20 sec
0 to 30 mph	4.7 sec
0 to 40 mph	7.1 sec
0 to 50 mph	9.8 sec
0 to 60 mph	14.1 sec
0 to 70 mph	22.1 sec
20 to 40 mph	4.1 sec
30 to 50 mph	5.05 sec
40 to 60 mph	7.8 sec

GO-TO-WHOA:

0 to 60 to 0 mph .... 18.4 sec

SPEEDO ERROR:

Indicated	Actual
30 mph	30.8 mph
50 mph	51.4 mph
70 mph	72.5 mph

FUEL CONSUMPTION:

21 mpg over 115 hard miles

PRICE:

(In Australia) £2150

TEST CAR FROM:

Preston Motors Ltd, Melbourne.

Nova SS
CHEVY II SPORT COUPE

MT Road Test

SS SERIES SPORTS NEW LUXURY OPTIONS FOR '63.

Lots of body lean and weird front-wheel angles are part of Nova's personality when cornering at 40 mph or more on smooth strip.

by Bob McVay, *Assistant Technical Editor*

THE CHEVY II NOVA 400-SS sport coupe boils down to a sales winner with excellent maneuverability, good operating economy, peppy performance, and a luxurious interior — all in a compact unit. This unit measures just 183 inches long, with a 110-inch wheelbase.

Having just stepped out of a full-sized, 2½-ton automobile, the small size and easy handling of the Chevy were quickly noticeable. This easy-to-get-into-and-out-of car really fills the bill for all those necessary short trips around town and to work and back. It's equally at home cruising on the open highway at legal speeds.

Our test car was Chevy II's top-of-the-line model. The SS, or Super Sport designation, located on the rear fenders and deck lid, tells the world that the car is equipped with those extra luxury options available only on the sport coupe and the convertible in the 400 series. These extras include a special instrument cluster that replaces the usual warning lights with gauges for oil pressure, ammeter, temperature, and fuel. Front bucket seats, all-vinyl interior, extra chrome trim both inside and out, and special wheel covers are included in the SS package.

A chrome transmission cover plate comes on models equipped with Powerglide, with the shift lever mounted on the floor. The SS accessory kit costs $161.40 extra. Fourteen-inch wheels and 6.50 x 14 tires are a required option on SS-equipped cars, but aren't part of the package. No engine modifications are included, but the six-cylinder unit is standard on the 400 series.

With the exception of a newly designed grille and trim changes (plus the amber turn-indicator lenses), the 1963 Nova continues with last year's successful styling. Registration records found Chevy II capturing 15.54 per cent of the compact market in 1962, placing it third behind Rambler and Falcon. December registrations found the "II" in front of the Falcon, so it looks like a close race for 1963.

One of the few compacts that hasn't gotten larger for 1963, the Chevy II gets its peppy performance from a six-cylinder, in-line engine of 194 cubic inches, rated at 120 hp at 4400 rpm. With a 3.563-inch bore and 3.25-inch stroke, the sturdy ohv Six gives 177 pounds-feet of torque at 2400 rpm.

It's an efficient unit, one of the Nova's good selling points, and during more than 800 miles of test driving, the lowest mileage we recorded was 16 mpg during heavy stop-and-go traffic. Out on the highway, cruising at 65-70 mph, the Chevy II gave 20.8 mpg without working hard. Comfortable highway speeds of 70 mph could be held for long distances without undue complaints or straining noises from the engine.

With the two-speed Powerglide transmission and standard 3.08-to-1 rear axle, the Nova surged off the line and accelerated to 30, 45, and 60 mph in 5.6, 9.5, and 15.9 seconds respectively, turning 67 mph in 21.0 seconds through the measured quarter-mile. Left in DRIVE all the way, the transmission shifted into high at 45 mph and 4300 rpm. Wound up tight, it hit a top speed of 91 mph while showing 4300 rpm on our electric tachometer. On all of our acceleration and high-speed runs we have two men and 100 pounds of test equipment aboard, so owners can expect slightly better times with only one person in the car.

Shipping weight of the sport coupe is 2605 pounds, but with a full tank of gas and the full complement of accessories, our test car weighed 2984 pounds (without test equipment). We found performance quite good for a car of this weight and with a single-barrel carburetor. After the transmission

CHEVY II NOVA

shifted into high at 45, acceleration dropped off somewhat, but there was enough power for reasonable passing safety in the 45-65-mph range.

The car buyer who doesn't take a close look at the list of options available for his car is doing himself and the car an injustice. There are many exceptional bargains available at extremely low prices when ordered with the car. These are installed at the factory. A quick look down the list shows the sharp-eyed buyer that for less than the price of a radio he can equip his Nova with the following heavy-duty equipment: radiator, front shock absorbers, springs front and rear, battery, and the highly desirable metallic brake linings for $52.85. An extra $37.70 buys the Positraction rear axle, and standard-shift cars can be ordered with a heavy-duty clutch for only $5.40 extra.

Under normal driving conditions at reasonable speeds, the Nova rides and handles well, giving a soft ride with very little vibration. Taking corners at higher-than-usual speeds brings out lots of body lean. There's some understeer, but it isn't uncomfortable in normal driving, and the excessive body lean on corners doesn't encourage high-speed cornering. Stiffer springs and shocks will help the car's handling and make it more comfortable to drive on winding roads. We found that when the car leaned too much, a decrease in speed solved the problem. It's a simple matter. This just isn't

a high-speed road car, but it does the job it was designed to do economically and well.

There's plenty of room for the usual odds and ends in the glove compartment, but even though the car has bucket seats and a long chrome transmission cover plate, no center console compartment is provided.

Adding a sporty flair to the interior, the Powerglide selector lever is mounted on the floor. Yet no light is provided. It's a matter of either learning the pattern or turning on the interior lights to see what range the transmission is in at night. This particular unit was vague in operation, allowing shifts directly from LOW into reverse, which isn't supposed to be possible and could be expensive if done at the wrong time. With the seat all the way forward, we had difficulty getting the lever into LOW range — it's all the way back and to the left on the quadrant. We feel the standard shift would be more desirable on the floor and the Powerglide selector should be back on the post with a lighted quadrant.

Six drivers put the Nova through its paces during our test period. This group included friends and their wives and, without exception, everyone complained about the seating arrangement. Moving the seat up so short drivers can reach

(ABOVE) *Independent front suspension contains shocks inside long, narrow coil springs, mounted on upper A-frames extending up into fender wells. Arrangement gives soft ride without front-end bottoming on sharp dips. Nine-inch brakes are power-assisted hydraulic units, give 145 squares inches' effective lining area. Metallic linings are very desirable option at only $37.70 extra.* **(LEFT)** *Wheel lock-up caused car to skid and swerve, making straight-line stops from 60 mph difficult. Pumping didn't help.*

the pedals brings them too close to the steering wheel. Moving the seat back away from the pedals makes it necessary for them to stretch. It seems to us that someone who pays $3000 (or even $2000) for a new automobile should be able to make it fit him comfortably. A seat that can be adjusted with simple tools to fit the size and shape of the driver would certainly be a welcome improvement. Some auto manufacturers already have such a seat. We hope others will follow.

Tall drivers didn't like the position of the parking brake handle. It's located under the dash to the right of the steering wheel post, and when pulled out it hit long-legged drivers' right kneecap.

The Chevy II uses independent front suspension, with double-acting shock absorbers mounted inside the front coil springs. The coils are attached to the upper A-frames and extend high into the front fender wells, allowing lots of wheel travel without front-end bottoming. A tension strut rod is attached to the lower control arm, and the weight of the front end rests on upper ball joints.

A new bearing surface, Teflon, is used on the upper ball joints and is the only lubricated point in the front suspension assembly. Teflon is claimed to extend lubrication periods from the usual 1000 miles to 6000 miles.

Three of the seven steering linkage points are now permanently sealed and lubricated, reducing the necessary lubrication fittings from five in 1962 to the current four.

Rear suspension has a semi-floating rear axle, with torque taken through two single-leaf springs of chrome carbon steel. Sedans and coupes use a 3.08-to-1 ratio with standard shift and automatic, while the four-cylinder station wagon uses a 3.55-to-1 ratio. The six-cylinder wagon has a 3.36-to-1 axle.
continued

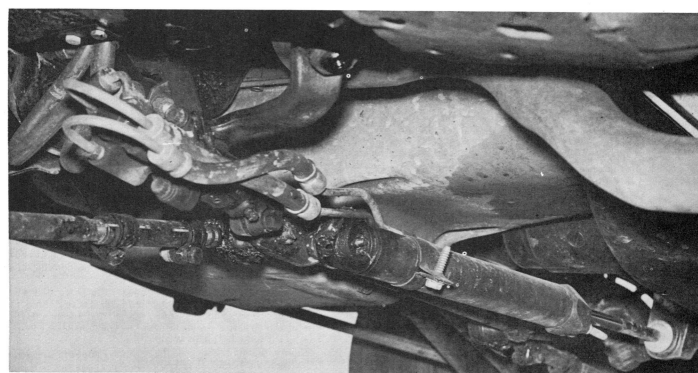
POWER-STEERING MECHANISM HANGS LOW UNDER CAR. HERE IT'S VULNERABLE TO ROCKS AND OTHER OBJECTS ENCOUNTERED DURING DRIVING.

AVERAGE STOPPING DISTANCE OF OUR LAST 18 TEST CARS WAS 153 FEET FROM 60 MILES PER HOUR. OUR NOVA REQUIRED 199 FEET.

CHEVY II NOVA

Standard tire size is 6.00 x 13 for sedans and 6.50 x 13 for the station wagons.

The 400 series offers the sport coupe, a convertible, a four-door sedan, and a four-door station wagon, with the six-cylinder engine as standard. The middle line is called the 300 series, and the least expensive line is the 100. Both lower series offer a two-door sedan, a four-door sedan, and a station wagon, with either two or three seats. Only the six-cylinder engine can be ordered with the three-seat station wagon, but either the Six or the Four can be ordered with the other body styles.

Also an ohv, in-line engine, the four-cylinder unit gets 90 horses from its 153 cubic inches and has a 3.875 bore, with a stroke of 3.25. It puts out 152 pounds-feet of torque at 2400 rpm and sells for just $59 less than the Six. We feel most buyers will want the extra punch offered by the Six, even though economy won't be so good as with the Four.

In addition to the two in-line engines, the performance-minded owner can drop one of the big Chevrolet V-8s into his Nova. The car was designed for this swap, with either the 340-hp, 327-inch engine or the 283-incher as dealer options. Performance on a 1962 Nova four-door sedan with the "327" engine, four-speed transmission, and Positraction resulted in 0-30, 0-45, and 0-60-mph times of 3.3, 5.0, and 6.8 seconds respectively.

Changing from the Six to the V-8 powerplant is expensive — the price depending on the individual dealer. We checked with one Los Angeles dealer and found that the necessary parts for the "327" conversion, with four-speed and Positraction, would cost $1555. This, however, doesn't include installation charges.

Our test car came equipped with the standard, self-adjusting nine-inch brakes with bonded asbestos linings. Holding the car in a straight line during our braking tests from 60 mph proved a real problem. Wheel lockup caused the car to skid and swerve, and even with lots of rapid pedal pumping the Nova covered 199 feet before coming to a complete stop from 60 mph.

On rough roads or smooth highways, the Nova is a well-behaved car when driven as it should be. There's room for five if necessary, and the trunk's large and roomy, having one of the easiest opening lids on any car. One twist of the key and it pops up all by itself.

Filling the gap between the Corvair and the full-sized Chevrolet, the Chevy II offers the compact-car buyer a high degree of comfort, economy, and reliability, and with the SS luxury equipment and the long list of options, should maintain its place at or very close to the top of the compact sales market. /MT

(LEFT) All shapes of luggage fit into Nova's roomy trunk. (BELOW) We liked the simple, easy-to-read gauges, with all controls in driver's reach.

The floor-mounted Powerglide lever was vague in operation and allowed shifts directly from LOW range to reverse, which isn't supposed to be possible. There's no quadrant light, so nighttime gear changes are pure guesswork unless you want to take the trouble to turn on interior lights.

CHEVY II NOVA 400-SS
2-door, 5-passenger hardtop

OPTIONS ON CAR TESTED: Powerglide, power brakes and steering, radio, padded dash, seat belts, whitewalls
BASIC PRICE: $2262
PRICE AS TESTED: $3056.70 (plus tax and license)
ODOMETER READING AT START OF TEST: 3283 miles
RECOMMENDED ENGINE RED LINE: 5200 rpm

PERFORMANCE
ACCELERATION (two aboard)
0-30 mph............................. 5.6 secs.
0-45 mph............................. 9.5
0-60 mph............................. 15.9
Standing start ¼-mile 21.0 secs. and 67 mph
Speeds in gears @ 4300 rpm
　Low ....45 mph
　High...............................91 mph (actual top speed)
Speedometer Error on Test Car
　Car's speedometer reading ......29　44　48　57　67　77
　Weston electric speedometer ....30　45　50　60　70　80
Observed miles per hour per 1000 rpm in top gear ....21 mph
Stopping Distances — from 30 mph, 45 ft.; from 60 mph, 199 ft.

SPECIFICATIONS FROM MANUFACTURER
Engine
Ohv, in-line 6
Bore: 3.563 ins.
Stroke: 3.25 ins.
Displacement: 194 cu. ins.
Compression ratio: 8.5:1
Horsepower: 120 @ 4400 rpm
Torque: 177 lbs.-ft. @ 2400 rpm
Horsepower per cubic inch: 0.61
Ignition: 12-volt coil

Gearbox
2-speed Powerglide; floor-mounted selector lever

Driveshaft
One piece — open tube

Differential
Hypoid — semi-floating
Standard ratio: 3.08:1

Suspension
Front: Independent, with bottom control arms, unequal-length upper A-arms, coil springs with integral shocks mounted above A-arms
Rear: Rigid axle; 2 single-leaf springs, tubular shocks; torque taken through leaf springs

Steering
Recirculating ball; power assisted
Turning diameter: 38.4 ft.
Turns: 4.5 lock to lock

Wheels and Tires
4-lug, steel disc wheels
6.50 x 14 2-ply tubeless tires

Brakes
Hydraulic, duo-servo, with power assist; cast-iron drums
Front: 9-in. dia x 2.25 ins. wide
Rear: 9-in. dia. x 1.75 ins. wide
Effective lining area: 145.0 sq. ins.

Body and Frame
Unitary construction; body-frame integral, bolt-on front end
Wheelbase: 110.0 ins.
Track: front, 56.8 ins.; rear, 56.3 ins.
Overall length: 183.0 ins.
Weight as tested: 2984 lbs. (full tank of gas)

THERE'S LOTS OF ROOM FOR SERVICING IN CHEVY II ENGINE ROOM, WHICH HOLDS ECONOMICAL 194-CUBIC-INCH, OHV, 120-HP SIX.

WIDE OPEN, NOVA 400-SS TURNED 91 MILES PER HOUR AND SHOWED RESPECTABLE QUARTER-MILE TIME OF 21.2 SECONDS, 67 MPH.

A 283-CU.-IN. V8 NEVER FOUND A HAPPIER HOME—We slung a big 195-hp 283-cubic-inch V8* into the Chevy II Nova Sport Coupe and now you'd think it was born that way.

This is the same Chevy II that spent a couple of happy years building up a following as one of the most wholesome things since brown bread. The one down-to-earth American car you wouldn't mind bringing home to mother or showing off to your friends. And the last car in the world you'd ever accuse of being pretentious. In short, a regular darb.

Now, with that V8 up front, Chevy II spends most of its time doing impressions of performance types. Give it a 4-speed all-synchro shift* and it's very close to being just that. After all, it started out with certain advantages: taut suspension, trim size, no-nonsense construction.

Is this any way for a nice, quiet, sturdy, sensible, unpretentious car like Chevy II to behave? Strangely enough, yes. Despite its new vigor, it's still a nice, quiet, sturdy, sensible, unpretentious car. With sharper teeth. Grrr.   **CHEVY II NOVA**   **CHEVROLET**

Chevrolet Division of General Motors, Detroit, Michigan   *Optional at extra cost

CHEVY II

Chevy II 400 Nova Series 4-door Saloon

ENGINE CAPACITY 194 cu in, 3,171.90 cu cm
FUEL CONSUMPTION 23.9 m/imp gal, 20 m/US gal, 11.8 l × 100 km
SEATS 6 **MAX SPEED** 97 mph, 156.2 km/h
PRICE $ 2,042

ENGINE front, 4 stroke; cylinders: 6, slanted at 3° 51', in line; bore and stroke: 56 × 3.25 in, 90.4 × 82.5 mm; engine capacity: 194 cu in, 3,171.90 cu cm; compression ratio: 8.5; max power (SAE): 120 hp at 4,400 rpm; max torque (SAE): 177 /ft, 24.4 kg/m at 2,400 rpm; max number of engine rpm: 4,800; specific power: 8 hp/l; cylinder block: cast iron; cylinder head: cast iron; crankshaft bearings: valves: 2 per cylinder, overhead, in line, slanted at 45°, push-rods and rockers, draulic tappets; camshaft: 1, side; lubrication: gear pump, full flow filter; lubricating system capacity: 8.27 imp pt, 10 US pt, 4.7 l; carburation: 1 Rochester 7023105 downdraught single barrel carburettor; fuel feed: mechanical pump; cooling system: water; cooling system capacity: 20.06 imp pt, 24 US pt, 11.4 l.
TRANSMISSION driving wheels: rear; clutch: single dry plate; gearbox: mechanical; gears: 3 + reverse; synchromesh gears: II, III; gearbox ratios: I 2.94, II 1.68, III 1, rev 2.94; gear lever: steering column; final drive: hypoid bevel; axle ratio: 3.08.
CHASSIS integral; front suspension: independent, wishbones, lower trailing links, coil springs, anti-roll bar, telescopic dampers; rear suspension: rigid axle, single leaf semi-elliptic springs, telescopic dampers.
STEERING recirculating ball; turns of steering wheel lock to lock: 4.50.
BRAKES drum, servo; braking surface: total 172.70 sq in, 1,113.91 sq cm.
ELECTRICAL EQUIPMENT voltage: 12 V; battery: 44 Ah; alternator; ignition distributor: Delco-Remy; headlights: 2.
DIMENSIONS AND WEIGHT wheel base: 110 in, 2,794 mm; front track: 56.80 in, 1,443 mm; rear track: 56.30 in, 1,430 mm; overall length: 182.90 in, 4,646 mm; overall width: 70.80 in, 1,798 mm; overall height: 55 in, 1,397 mm; ground clearance: 5.20 in, 132 mm; dry weight: 2,740 lb, 1,243 kg; distribution of weight: 53.5% front axle, 46.5% rear axle; turning circle (between walls): 39.5 ft, 12 m; width of rims: 4''; tyres: 6.00 × 13; fuel tank capacity: 13.4 imp gal, 16 US gal, 61 l.
BODY saloon; doors: 4; seats: 6; front seats: bench.
PERFORMANCE max speeds: 36 mph, 58 km/h in 1st gear; 62 mph, 99.8 km/h in 2nd gear; 97 mph, 156.2 km/h in 3rd gear; power-weight ratio: 22.9 lb/hp, 10.4 kg/hp; carrying capacity: 1,058 lb, 480 kg; speed in direct drive at 1,000 rpm: 21.8 mph, 35.1 km/h.
PRACTICAL INSTRUCTIONS fuel: 91-94 oct petrol; engine sump oil: 6.69 imp pt, 8 US pt, 3.8 l, SAE 5W-20 (winter) 10W-30 (summer), change every 6,000 miles, 9,700 km; gearbox oil: 1.58 imp pt, 2 US pt, 0.9 l; final drive oil: 2.99 imp pt, 3.50 US pt, 1.7 l; greasing: every 6,000 miles, 9,700 km, 9 points; valve timing: inlet opens 34° before tdc and closes 86° after bdc, exhaust opens 68° before bdc and closes 52° after tdc; tyre pressure (medium load): front 24 psi, 1.7 atm, rear 24 psi, 1.7 atm.
VARIATIONS AND OPTIONAL ACCESSORIES limited slip final drive; power-assisted steering; heavy-duty elements; 6.50/7.00 × 13 and 6.50 × 14 tyres; 3.36 3.55 axle ratios; Powerglide automatic gearbox, hydraulic torque convertor and planetary gears with 2 ratios (I 1.82, II 1, rev 1.82), 3.08 axle ratio; V8 engine, capacity 283 cu in, 4,627.05 cu cm, 195 hp at 4,800 rpm, 3-speed mechanical gearbox (I 2.58, II 1.48, III 1, rev 2.58), 2.58 3.08 3.36 axle ratios, 4-speed mechanical gearbox (I 2.56, II 1.91, III 1.48, IV 1, rev 2.64), 2.56 3.08 3.55 axle ratios, automatic gearbox, max speed 110 mph, 177.1 km/h; export model, 110 hp at 4,400 rpm; Chevy II 100 Standard Series 2-door Saloon, only with engine capacity 153 cu in, 2,501.55 cu cm, 90 hp at 4,000 rpm, max speed 86 mph, 138.5 km/h; Chevy II 100 Standard Series 4-door Saloon; Chevy II 100 Standard Series Estate Car, only with engine capacity 194 cu in, 3,171.90 cu cm, max speed 95 mph, 152.9 km/h; Chevy II 200 Standard Series 2-door Saloon; Chevy II 200 Standard Series 4-door Saloon; Chevy II 200 Standard Series Estate Car; Chevy II 400 Nova Series 2-door Saloon; Chevy II 400 Nova Series Estate Car.

by Bob McVay
Assistant Technical Editor

CHEVY II V-8 ROAD TEST

LAST YEAR, you couldn't order a Chevy II V-8 from the factory, but you could have one installed by your friendly dealer for $1500 plus labor. Getting a V-8 Chevy II this year is a lot less trouble and *much* less expensive. Option L-32 plus an extra $107.60 brings a new Chevy II into town with a 283-cubic-inch V-8 neatly installed, without any additional labor charges or extra waiting.

Lots of things have been going on at the Chevy II works lately. First, they decided to drop the Nova SS convertible and hardtop models, hoping people who'd normally buy them would upgrade themselves to similar models of the new Chevelle. Such wasn't the case, and soon the howls of customers and dealers alike brought a re-introduction of the popular hardtop model. The convertible has yet to make its reappearance.

The big change for 1964 is the new engine option and a slight increase in the size of the Chevy II's brakes, plus a few other changes. Except for some emblem and trim relocations, this year's Chevy II hasn't changed much in appearance. Biggest difference former Chevy II owners will notice is the 75-hp increase. Last year's Six wasn't a neck-snapper, but the 195-hp V-8 is. It takes the Chevy II out of the ho-hum category and makes it fun to drive — at least a lot more fun than the 120-hp Six version.

Fully equipped with every possible power accessory and comfort option, our test hardtop weighed 3000 pounds ready for action. Our (standard) 3.08 rear axle wasn't the best choice for performance, but it's a good one for overall family driving. Even so, it peeled nearly five seconds off last year's 0-60-mph time, with a best run of 11.3 seconds. Thirty and 45 came up in 3.5 and 6.8 seconds, but when the two-speed automatic made its first and only shift at 51 mph and 4300 rpm, acceleration dropped off considerably until the car expired completely at 4500 rpm and an honest 100 mph. Our measured quarter-mile

CHEVY II STILL LEANS QUITE A BIT AND UNDERSTEERS IN CORNERS, BUT OPTIONAL V-8'S ADDED POWER MAKES FOR MUCH BETTER CONTROL.

came up in 18 seconds flat, with a top speed of 75 mph.

Optional ratios of 3.36 and 3.55 are available for mountain driving and better acceleration. Either the standard three-speed column shift or Chevrolet's new four-speed Muncie gearbox option ($188.30) would knock seconds off the car's acceleration times, since Powerglide isn't the best choice for top performance. The "283" is also Chevelle's top engine, and with a four-barrel carb giving 220 horses, it shouldn't take a great deal of mental gymnastics or money to make a real performer out of the Chevy II.

A Chevy II sedan, equipped with the V-8, starts for as little as $2167.60 with standard three-speed transmission. It'd be a lot lighter than our test car, because it wouldn't be loaded with heavy, power-stealing accessories. Its performance should be much better than our test car's due to weight saving alone. Now add Positraction, that nice, precise, four-speed transmission, and one of the optional axle ratios — and you've got a Chevy II that goes. Since the "283" V-8 has been around for years, speed shops and parts houses carry a huge array of bolt-on options and special equipment for it. Now that the price of a factory-installed V-8 option is within reason, its possibilities are almost limitless for the do-it-yourselfer.

In addition to engine possibilities, heavy-duty options can be ordered to make the car handle and stop better than in standard form. One particularly helpful option would be the sintered metallic brake linings that go for a modest $37.70, plus Positraction at the same price. By ordering the options and accessories carefully (the car's light enough so it doesn't really need power brakes or steering, especially since power steering doesn't decrease the turns between locks), you could have a Chevy II V-8 with exceptionally good performance capabilities for around $2600.

So much for performance talk. The Chevy II is primarily an economy compact and fits that niche very nicely. Its light handling (if somewhat vague with power assist) and small exterior dimensions have endeared it to thousands of wives and husbands. It's an exceptionally good choice for an around-town car, because it's easy to maneuver and park and has a short, flat hood for good all-around vision.

In addition to its easy handling and peppy performance, it has other endearing features that make it a good family car. It has a big trunk for its size, the lid pops open with a twist of the key, and women will appreciate the all-vinyl interior that's durable and easy to keep clean. Station wagons and SS models have all-vinyl interiors, while sedans use

New grille, lack of ornamentation give Chevy II clean, uncluttered look.

CHEVY II

Accessory-loaded car with automatic was no drag strip contender, but one with four-speed trans, Positraction, and a few other mods could be fun on street or drag strip.

PHOTOS BY BOB D'OLIVO

Only two-throat carb is offered with V-8 package, but bolt-on goodies are readily available for this popular engine. Chevy's "283" has been on the market since '57.

Dash of SS model is simple, easy to read, and uses gauges instead of warning lights. Automatic shift lever is vague in operation. Protruding knobs could bark knuckles.

a combination of cloth with vinyl trim in five color selections: fawn, aqua, red, blue (ours was blue), and saddle.

The Chevy II's seating still leaves something to be desired. Those who tried it felt the seat backs didn't have enough rake and that some means of seat adjustment should be provided so owners can make their cars fit them in each individual case.

Out on the road, the Chevy II has adequate handling for normal driving. Chasing sports cars over winding, mountain roads isn't its forte. If not exciting, its handling is honest and predictable. The heavier V-8 makes itself felt during cornering. The car understeers — more as the speed gets higher — and the car isn't at its best in a fast bend, although the extra 75 horses give a greater amount of control for holding a line through a turn. Excessive body lean is still a characteristic.

We found the car much more pleasant at average cruising speeds. Body lean, understeer, and tire scrubbing telegraph the car's limits to the driver. The Nova didn't break loose suddenly on dry pavement, but once the road got a little damp, the forward weight bias made itself felt. You have to take extra care to avoid wheelspin on take-off and skidding when making tight turns on slick streets.

Neither does the Chevy II lend itself to overly high-speed driving. It's perfectly content cruising at any legal speed, but the front end feels progressively lighter as speed increases, and crosswinds have a tendency to change the car's direction at higher speeds.

Our test car seemed well built and well detailed inside and out. Wind noise wasn't excessive at normal cruising speeds, but the commotion caused by full-speed operation of the heater or air conditioner was enough to raise conversation to shouting. Both units did their jobs well and, once the desired temperature was reached, the blowers could be turned down to maintain the temperature.

Ride wasn't bad, but the shorter-wheelbase Chevy II tends to ride somewhat rougher than its longer-wheelbase stablemates. The heavier V-8 helps hold the front end down so it doesn't bounce much on sharp dips, and the suspension soaks up all but the sharpest bumps

---

(OPPOSITE) *Chevy II was never meant as a road racer, but handling is predictable on winding roads. Understeer and body lean increase with speed, letting the driver know just how far away he is from the car's limit. Car rides well, is comfortable.*

CHEVY II

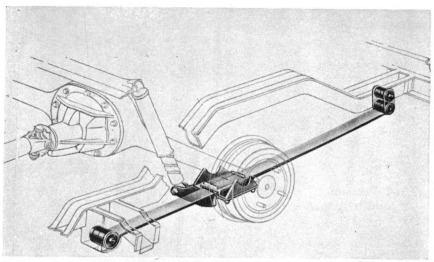
Unusual single-leaf rear springs, rubber-bushed at shackle and hanger, are made of chrome carbon steel, have identical spring rates as six-cylinder Chevy II offerings.

SLA front suspension uses high-mounted coils extending into fender well. The power-steering mechanism still hangs too low — it could easily be damaged by obstructions.

Car has very roomy trunk for its size, allows two-suiter to stand up and spare to be removed without unloading everything. Lip is low for easy loading, lid opens wide.

well enough for a reasonably comfortable ride.

The Chevy II's power brakes were a bit touchy in operation, but they're lots stronger stoppers than last year's. Effective lining area is increased by 27.7 square inches, and drum diameter has gained a half inch. Wheels tended to lock up fairly quickly on panic stops, but the car didn't swerve too much, and it wasn't hard to keep in a straight line. Average compared with other 1964 cars, Chevy II's brakes are a big improvement over those of last year. They gave stops from 60 mph in 169 feet — that's 30 feet less than our 1963 Chevy II took.

For those who put economy above all else, Chevy II still offers their in-line Four, rated at 90 hp. It's standard with a three-speed manual transmission and optional with Powerglide, as is the 120-hp Six. Somehow, we feel lots of buyers are going to prefer the V-8 and will be quite willing to pay the extra dollars for the extra performance. We certainly would.

Additional models in the Chevy II line-up include two- and four-door sedans and a four-door, six-passenger station wagon in the Nova and lower-priced "100" series. The standard Sport Coupe is offered in the Nova series with bench seats. The Super Sport hardtop (our test car) tops the line. Who knows? If buyers scream long and loudly enough, maybe they'll even bring back that nice little convertible. It'd make a fine package with the V-8 option.

Loaded with every possible option or in stripped form, the Chevy II makes sense as a family car. It's easy to drive and meets most of the needs of normal-sized families. It also shares Chevrolet's 24-month/24,000-mile warranty as well as their extended maintenance periods. It's more a daily go-to-work car than a high-speed ground-eater, but it's equally at home in traffic or on the highway at legal speeds. Its size and maneuverability make it a good choice for those who do most of their driving in traffic, and any of its three engines should give long life and good fuel economy — they're all designed to run on regular fuel. As an example, our test car with automatic transmission gave a high of 19.6 mpg on the highway and a 12.3 low in heavy traffic, with an overall average of 14.4 mpg.

The V-8 is the big news for 1964. Now that the price is within reason, it can range from a family workhorse to a low-priced hot rod, depending on options and accessories. By adding a V-8 and bigger brakes, plus detail changes, Chevrolet has made a nice compact even more desirable and a much better performer. /MT

ALTHOUGH CAR DOESN'T HAVE OUTSTANDING STOPPING POWER, BIGGER BRAKES SHORTENED LAST YEAR'S BRAKING DISTANCES BY 30 FEET.

CHEVY II NOVA SS SPORT COUPE
2-door, 5-passenger hardtop

OPTIONS ON CAR TESTED: Powerglide, air conditioning, power brakes and steering, radio with rear-seat speaker, heater, tinted glass, whitewalls, seat belts, misc. access.
BASIC PRICE: $2550.75
PRICE AS TESTED: $3503.15 (plus tax and license)
ODOMETER READING AT START OF TEST: 1200 miles
RECOMMENDED ENGINE RED LINE: 4800 rpm.

PERFORMANCE

ACCELERATION (2 aboard)
```
0-30 mph............................. 3.5 secs.
0-45 mph............................. 6.8
0-60 mph.............................11.3
```
Standing start ¼-mile 18.0 secs. and 75 mph
Speeds in gears @ shift point
```
    1st..............51 mph @ 4300 rpm
    2nd..............100 mph (actual top speed) @ 4500 rpm
```
Speedometer Error on Test Car
Car's speedometer reading......32	48	54	64	74	84
Weston electric speedometer....30	45	50	60	70	80

Observed miles per hour per 1000 rpm in top gear.............23.0 mph
Stopping Distances — from 30 mph, 30 ft.; from 60 mph, 169 ft.

SPECIFICATIONS FROM MANUFACTURER

Engine
Ohv V-8
Bore: 3.875 ins.
Stroke: 3.00 ins.
Displacement: 283 cu. ins.
Compression ratio: 9.25:1
Horsepower: 195 @ 4800 rpm
Torque: 285 lbs.-ft. @ 2400 rpm
Horsepower per cubic inch: 0.69
Carburetion: 1 2-bbl.
Ignition: 12-volt coil

Gearbox
2-speed automatic (Powerglide); floor-mounted lever

Driveshaft
1-piece, open tube

Differential
Hypoid, semi-floating
Standard ratio: 3.08:1

Suspension
Front: Independent, SLA, high-mounted coil springs, with integral double-acting tubular shocks, anti-sway bar
Rear: Rigid axle, with single-leaf semi-elliptic springs, double-acting tubular shocks

Steering
Recirculating ball nut, with linkage-type power assist
Turning diameter: 38.4 ft.
Turns lock to lock: 4.5

Wheels and Tires
5-lug, steel disc wheels
6.50 x 14 4-ply rayon whitewall tires

Brakes
Hydraulic, duo-servo; cast-iron rim, steel web, with integral power assist; self-adjusting
Front: 9.5-in. dia. x 2.5 ins. wide
Rear: 9.5-in. dia. x 2.0 ins. wide
Effective lining area: 172.7 sq. ins.

Body and Frame
Unit construction
Wheelbase: 110.0 ins.
Track: front, 56.8 ins.; rear, 56.3 ins.
Overall length: 182.9 ins.
Overall width: 70.8 ins.
Curb weight: 3000 lbs.

CAR LIFE ROAD TEST

CHEVY II 327/350 V-8
Who Needs 400 Inches for Supercar Status?

A LOT of Chevy IIs have gone under the overpass since CAR LIFE acclaimed the initial model with our 1962 Award for Engineering Excellence. The Award, to some readers, may have seemed curious since the Chevy II had few major components and no structural members which could be regarded as novel. It goes without saying, however, that novelty in itself is no criterion of excellence—or even adequacy—in engineering. What justified the Award was the highly competent manner in which the Chevy II was engineered to meet certain cost and weight standards for a specific transportation need. At the time, it will be recalled, the car was available with either a 153-cu. in. 4-cyl. or 194-cu. in. 6-cyl. engine.

The following year, in 1963, the CL Award was presented to the Corvette Sting Ray, for entirely different reasons. Indeed, one may say that the engineering considerations involved in the Sting Ray case were exactly opposite: Cost and price were not the object; rather, the goal was the engineering of an "ultimate" passenger vehicle. Since this period preceded the appearance of the latest 396/427-cu. in. engine, the powerplant involved was the 327-cu. in. V-8 which has come to be regarded as the "Corvette engine."

Now Chevrolet Motor Division has brought together elements of the two Award winners, mating the 327 engine with the Chevy II. The result is a regular production version of an interesting hybrid built up a few years ago by race tuner Bill Thomas. Although his car, christened "Bad Bascomb," went one step farther and utilized the Sting Ray rear suspension, the 1966 Nova SS retains the Chevy II single-leaf rear springs and Hotchkiss drive. An even earlier Thomas car which was the subject of a full-scale road test (CL, June '63) was somewhat less ambitious in that it only consisted of an engine swap; but, because it utilized a fuel-injected 327 Corvette engine, perhaps it should be considered a prototype for this year's production car.

As a matter of fact, Chevrolet has regularly increased the engine power options available for the Chevy II almost since the car's inception. For two years now, the 283-cu. in. V-8 has been readily available. And, because the 283 and 327 are based on a similar block, the Corvette engine option has been only logical. Buyers are demanding more and more power in their cars, not only as power for power's sake,

HEART of the hustle is 350-bhp Corvette V-8 tucked under the hood.

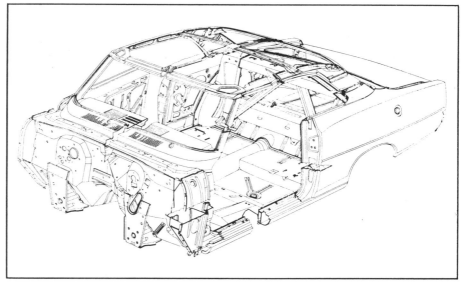

MUCH OF THE impressive solidity of the test car is due to the sturdy unitized body shell; a separate stub frame carries engine and front suspension. Body is rattle-free.

RACING WHEEL option would seem more at home in the business office.

BENEATH THAT mild-mannered disguise it's really Supercar.

but also as a means to operate the proliferating array of comfort and convenience options. Chevrolet has kept its Chevy II in a position to answer these market demands, despite the fact that the car originally was designed for the frill-free, economical, compact and minimal transportation needs of a more stringent economy. The Chevy II, with its original engines still offered, stands without peer when it comes to variety in choice of power.

THE 327 ENGINE is offered in two power ratings, 275 and 350 bhp, with more than a few distinguishing characteristics between them. The test car was equipped with the more powerful engine, which, like oysters on the half shell, provides a pleasant surprise every time it is tried. There is no doubt that this is a strong engine, indeed, but it also is as tractable as an economy Six. When producing its full output, it contributes a most interesting power/weight ratio for the 3200-lb. car. Yet tiddling around in town traffic shows a gentle and docile side of the engine that is remarkable for such a high-performance, profoundly-carburetored prime mover.

Nothing really unusual is done to the engine to make it perform so well, but what is done testifies to a great deal of talent on the factory front. There is a high compression (11.0:1), a generous carburetor (Holley 4-barrel with 1.561-in. barrels) and unrestricted fuel/air charge passages (2.02-in. intake valves, 1.60-in. exhausts). However, its hydraulic lifters obviously have a faster bleed-down rate than any of the competitors' and its camshaft has to represent some sort of pinnacle in precision design. Though remarkably long for a production car engine, the camshaft duration is not completely what the listed 342° duration would lead one to believe since Chevrolet traditionally includes ramps in the figure. An overlap of 114° and a high, 0.447 in., lift guarantee the engine's ability to keep the gas mixture moving well into the higher rpm ranges. It should be noted that a more mild camshaft, listed at 300° with 78° overlap and a lesser, 0.3987 in., lift, is used in the 275-bhp version of the engine.

The engine provides surprisingly strong punch up to 4000 rpm, all through the lower rpm range where such powerplants are normally expected to be lightweight performers. Above that point, it's like having another, even larger, engine suddenly switched on. In view of the power/weight ratio and the engine's performance characteristics, acceleration sprints are of a nature to satisfy the most demanding enthusiast or frighten the less sure. It is interesting to note that the test car's performance on the quarter-mile was very close to that of the '62 327 Chevy II, as are the output ratings of the present engine and the 360-bhp fuel-injected version then installed. This production car did have the next higher final drive ratio, however.

Ordinarily, the Muncie 4-speed manual transmission proves to be an exceptional gearbox which can be used with so little effort that one tends to overlook it. Yet, for an inexplicable reason, the unit in the test car did not live up to the reputation of past examples. It was difficult to shift, requiring more than a little exertion, and gears tended to be hard to "find." The problem may well have been eliminated by some sort of service adjustment, but it was one the test drivers had never before experienced with the transmission. In addition, the spacing in the 2.52-series gearset leaves a gap between second and third that is particularly evident when maximum performance is sought. The really serious enthusiast should specify the optional Muncie gearbox with its better spaced 2.20-series gearset with the 350-bhp engine. Moreover, in most easy driving situations the CL test crew took advantage of the engine's power to shift directly from first into third or high, utilizing the engine's surplus of power for brisk acceleration.

Almost as fascinating as the engine performance are the handling qualities of the car, despite the increased front-end heaviness which is the common penalty of a V-8 over a Six. The car could be cornered with a great deal of

CHEVY II

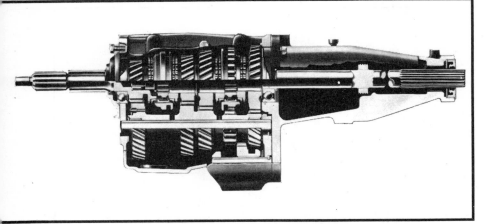

FULL-SYNCHRO Muncie 4-speed gearbox is well regarded but suffered in test car from an extremely stiff shift linkage. Optional close-ratio gearsets are available.

unconcern—the limiting factor appeared to be the grip of the tires. True to their advertised claim, the U.S. Royal Laredo tires did exhibit better-than-usual adhesiveness on wet pavement although performance on dry surfaces seemed no more than ordinary. Larger size tires specified with the 327 engine option, of course, contributed something to the handling. Stiffer spring rates also are part of the package, as is a 0.87-in. anti-roll bar in the front. The latter was a source of annoying and expensive-sounding noises on the test car, but examination disclosed that the bar's rubber bushings merely had shredded away before the car had seen 1000 miles of service. In all, while its weight distribution is in keeping with contemporary practice, the level of handling was a cut above competitive. Suspension control was well damped and without vice, something totally unexpected in view of the single-leaf rear springs.

Any evaluation of handling—or riding quality, for that matter—must include mention of the special seats fitted to the Chevy II Nova. These are much like those in the deluxe Impala or Caprice and have an unusually high back topped with a stylish triangular headrest. A departure from normal GM bucket seat practice in that they have definite side bolsters on the seat cushion, they provide quite good lateral support.

WEAK SPOTS in the car, as might be expected, are in the steering and the brakes. With power assist, the steering requires no effort and could (but doesn't) have a quicker overall ratio. As it was, the 4.5 turns between locks was too slow for the type of driving this car inspires. As for the brakes, they, too, are a familiar problem. Despite all the engine power (and weight)

1966 CHEVY II
NOVA SS 327 HARDTOP

DIMENSIONS
Wheelbase, in.	110.0
Track, f/r, in.	56.8/56.3
Overall length, in.	183.0
width	71.3
height	53.8
Front seat hip room, in.	2 x 23.3
shoulder room	55.3
head room	37.2
pedal-seatback, max.	44.0
Rear seat hip room, in.	58.6
shoulder room	53.8
leg room	31.0
head room	36.4
Door opening width, in.	43.6
Floor to ground height, in.	11.0
Ground clearance, in.	6.4

PRICES
List, fob factory	$2480
Equipped as tested	3662
Options included: 327/350 V-8, 4-speed trans., limited-slip diff., power steering, power brakes, H.D. suspension, air cond., emission control, smog valve, radio and rear speaker, deluxe bucket seats, belts and console, tinted windows, instruments.	

CAPACITIES
No. of passengers	5
Luggage space, cu. ft.	13.0
Fuel tank, gal.	16.0
Crankcase, qt.	5.0
Transmission/diff., pt.	3/3.5
Radiator coolant, qt.	17.0

CHASSIS/SUSPENSION
Frame type	unitized
Front suspension type: Independent, s.l.a. with coil springs and concentric telescopic shock absorbers, ball joint steering knuckles.	
ride rate at wheel, lb./in.	101
anti-roll bar dia., in.	0.867
Rear suspension type: Live axle, single leaf semi-elliptic springs, telescopic shock absorbers.	
ride rate at wheel, lb./in.	121
Steering system: Linkage assisted, semi-reversible ball nut, parallelogram linkage.	
gear ratio	20.0
overall ratio	25.4
turns, lock to lock	4.5
turning circle, ft. curb-curb	38.4
Curb weight, lb	3140
Test weight	3530
Weight distribution, % f/r	54.5/45.5

BRAKES
Type: Single-line hydraulic duo-servo, self-adjusting shoes in composite drums.	
Front drum, dia. x width, in.	9.5 x 2.5
Rear drum, dia. x width	9.5 x 2.0
total swept area, sq. in.	268.6
Power assist	integral, vacuum
line psi @ 100 lb. pedal	815

WHEELS/TIRES
Wheel size	14 x 5J
optional size available	13 x 4J
bolt no./circle dia., in.	5/4.75
Tire make, brand	U.S. Royal Laredo
size	6.95-14
recommended inflation, psi	24
capacity rating, total lb	3680

ENGINE
Type, no. cyl.	V-8, ohv
Bore x stroke, in.	4.00 x 3.25
Displacement, cu. in.	327
Compression ratio	11.0
Rated bhp @ rpm	350 @ 5800
equivalent mph	123
Rated torque @ rpm	360 @ 3600
equivalent mph	79
Carburetion	Holley, 1 x 4
barrel dia., pri./sec.	1.561
Valve operation: Hydraulic lifters, pushrods and rocker arms.	
valve dia., int./exh.	2.02/1.60
lift, int./exh.	0.447
timing, deg.	54-108, 102-60
duration, int./exh.	342
opening overlap	114
Exhaust system: Dual, reverse flow mufflers with resonators.	
pipe dia., exh./tail	2.5/2.0
Lubrication pump type	gear
normal press. @ rpm	30-45 @ 1500
Electrical supply	alternator
ampere rating	37
Battery, plates/amp. rating	61/20

DRIVE-TRAIN
Clutch type: single centrifugal dry disc	
dia., in.	10.4
Transmission type: Manual, 4-speed.	
Gear ratio 4th (1.00) overall	3.31
3rd (1.46)	4.83
2nd (1.88)	6.22
1st (2.52)	8.34
1st x t.c. stall ( )	
synchronous meshing?	all
Shift lever location	console
Differential type: Semi-floating with overhung pinion	
axle ratio	3.31

CAR LIFE

increases, the 9.5-in. drum brakes are still the same units initially supplied for the 4-cyl. Chevy II. A vacuum power booster in the test car was virtually unmanageable during the all-on stops from 80 mph, locking up any or all wheels at unpredictable times and causing an embarrassing loss of control.

The slow steering ratio, moreover, compounded the problem by denying the driver any ability to properly (and instantly) correct for such behavior. So long as Chevrolet borrows the Sting Ray's engine, it should also include that car's disc brakes and quick steering arms to complete the project.

One of the least kept secrets in Detroit is that the Chevy II is, for all practical purposes, the mechanical prototype for Chevrolet's forthcoming Panther (*CL*, May '65). The test car provides vivid proof that this is a sound basis upon which to build a sporting type of car. The Chevy II 327, for example, could hold its own among all Mustangs except the Shelby GT-350. Its Corvette engine puts all lesser Mustangs in the shade, although its lack of effective brakes leaves an important plus in the opposition camp's disc brake option. Those who simply can't wait for the Panther could have

LARGE TRUNK for size of car is one distinct advantage of front-engined, rear-drive layout. It will hold a family-sized quantity of luggage.

all but the highly styled exterior by opting for the present Chevy II Nova.

More intriguing, however, is the fact that the Chevy II 327 relates more to the present proliferation of Supercars than it does to a counter-Mustang. And in that context, it is well worth a close examination. Unlike some samples from the Supercar spectrum, it maintains a gentleness along with its fierce performance potential; its power/weight ratio is second to none and it is definitely better balanced than most. While admittedly giving a cubic inch advantage away to the more established models, the Corvette engine manages to be just as competitive in pure output. On the basis of specific bhp/cu. in. ratios, as a matter of fact, it stands heads above the Supercar level. ∎

CAR LIFE ROAD TEST

CALCULATED DATA	
Lb./bhp (test weight)	10.1
Cu. ft./ton mile	115
Mph/1000 rpm (high gear)	21.2
Engine revs/mile (60 mph)	2830
Piston travel, ft./mile	1532
Car Life wear index	43.4
Frontal area, sq. ft.	21.3
Box volume, cu. ft.	406.5

SPEEDOMETER ERROR	
30 mph, actual	30.2
40 mph	40.6
50 mph	50.1
60 mph	59.2
70 mph	68.2
80 mph	77.6
90 mph	86.6

MAINTENANCE INTERVALS	
Oil change, engine, miles	6000
transmission/differential	6000
Oil filter change	6000
Air cleaner service, mo.	6
Chassis lubrication	36,000
Wheelbearing re-packing	6000
Universal joint service	none
Coolant change, mo.	24

TUNE-UP DATA	
Spark plugs	AC-44
gap, in.	0.035
Spark setting, deg./idle rpm.	10/900
cent. max. adv., deg./rpm.	30/5100
vac. max. adv., deg./in. Hg.	15/12
Breaker gap, in.	0.019
cam dwell angle	28-32
arm tension, oz	19-23
Tappet clearance, int./exh.	0/0
Fuel pump pressure, psi.	5-6.5
Radiator cap relief press., psi.	15

PERFORMANCE	
Top speed (5800), mph	123
Shifts (rpm) @ mph	
3rd to 4th (5800)	85
2nd to 3rd (5800)	63
1st to 2nd (5800)	49

ACCELERATION	
0-30 mph, sec.	2.6
0-40 mph	3.7
0-50 mph	5.5
0-60 mph	7.2
0-70 mph	9.2
0-80 mph	11.3
0-90 mph	14.2
0-100 mph	18.2
Standing ¼-mile, sec.	15.1
speed at end, mph	93
Passing, 30-70 mph, sec.	6.6

BRAKING	
(Maximum deceleration rate achieved from 80 mph)	
1st stop, ft./sec./sec.	22
fade evident?	yes
2nd stop, ft./sec./sec.	20
fade evident?	yes

FUEL CONSUMPTION	
Test conditions, mpg	13.5
Normal cond., mpg	12-15
Cruising range, miles	192-230

GRADABILITY	
4th, % grade @ mph	18 @ 87
3rd	25 @ 73
2nd	31 @ 62
1st	40 @ 42

DRAG FACTOR	
Total drag @ 60 mph, lb.	124

ACCELERATION & COASTING

ELAPSED TIME IN SECONDS

MAY 1966

CHEVY II NOVA 327 ROAD TEST

BOTH THE CHEVY II and the 327 V-8 were introduced in 1962, but it wasn't until 1965 that they met on the assembly line. The first 327-cubic-inch V-8 offered in the Chevy II was a rather conventional 250-hp version, which has been upped to 275 for the current model run. Along with the 275-hp engine, Chevrolet, because of the good response from buyers of the 327 option, has made available a 350-hp option of the same displacement. However, we felt that the 275-horse Nova was the more interesting fare for the reader with a family and an eye for some compromise between economy and performance.

Our test car was a Nova Super Sport 4-passenger sport coupe. Besides the 327 V-8, it was equipped with Powerglide transmission, power steering, power brakes, radio and whitewall tires — all optional equipment. The car weighed in at 3040 pounds, which is light for any car so equipped, and undoubtedly had a lot to do with our surprising mileage figures.

We recorded a best of 19.2 mpg. This figure wasn't obtained on a long-distance trip, either, but rather under typical urban freeway conditions. The poorest mileage was 15.2 mpg, encountered during around-town driving, complete with traffic tie-ups. Such mileage is a real plus in the Chevy II's favor. We attribute low fuel consumption to the good power-to-weight ratio of 11.4 pounds per horsepower, making the engine's job an easy one.

Despite the new outside sheetmetal and minor changes inside, the Chevy II didn't feel vastly different from earlier models. Driving it, you feel as if you are guiding a single piece of metal that will go where you want it to, when you want it to. It's a good, controllable automobile. Driver positioning behind the 16.24-inch steering wheel is excellent, though with the "Astro" bucket seats, which our car had, a 6-plus-footer has to be careful of his head touching the headliner. Definitely, he couldn't wear a hat.

All-around vision isn't bad, but the large rear-view mirror takes a little getting used to. It's suspended from the top of the windshield frame and is positioned almost midway between the dash and roof. This allows very good vision to the rear but steals a bit from the forward vista. We learned to compensate for it, but were never completely happy with the positioning.

Style-wise, the Nova SS isn't likely to win any design laurels, but neither is it a bad-looking car. The plain-Jane-ness of its appearance has to grow on you and after it does, you tend to favor it. The new fender panels do make the car look bulkier than before, certainly a trend in tune with the times.

We were pleasantly surprised at the size of the luggage compartment. Our full complement of test equipment, a bulky load that some "compacts" just won't handle, fitted in easily. The lift-over height is under two feet, and the counterbalanced deck lid shuts without slamming.

The dashboard has a clean, functional look to it, and all the switches and gauges (what gauges there were, that is) are well placed. The heater controls are at the mid-section directly under the radio and don't require a master's degree to decipher. We found the "across-the-board" speedometer to be consistently accurate, being three mph fast all the way. This is the approximate boost which all car makers endeavor to build into the instrument, but this particular one was consistent.

The shift lever for the console-mounted linkage has a push bar located on the top of the "T" handle. We'd prefer a bar further around the handle that could be depressed by the fingers rather than the palm of the hand. Fingers seem to naturally fall around the "T" handle, but when you have to push down with your palm, then pull the selector into gear, it's an unnatural process.

Also, 2-stop doors would be a welcome improvement. The single stop now used has an exaggerated eccentric action, and unless you keep your hand on the door when you open it, it quickly swings out into the fully open position. This is a built-in source of dented, chipped or scratched doors, for others as well as the Chevy II owner. These complaints are not serious, just food for thought by the manufacturer.

Assembly plants, we find, *can* goof. The Chevy II was running well during our tests at Carlsbad Raceway, and with each acceleration run seemed to be bettering itself. But after our third quarter-mile pass, we noticed water shooting from between the hood and fender line. Thinking that it might be water from the overflow pipe being blown around by the fan, we waited a minute, watching from inside the car. When it didn't slow down or quit, we jumped out, pulled the hood open and discovered the upper radiator hose had a hole sawed clean through it. Water was pouring onto the alternator cooling fins, which in turn were spinning the water out toward the hood and fender. Investigating further, we found that upon acceleration, when the torque of the engine caused it to move down on the right and up on the left, the alternator's fins would come in contact with the hose and bite it, with a highly deleterious effect.

Don't think, though, that this is a common problem with 327-equipped Chevy IIs — just ours! It seems that the wrong radiator hose was put on at the assembly plant, and no one had

Neat appearance of rear end should stay that way with the protection of large bumper, warding off all but the tallest of trucks. Large rear glass aids vision of driver behind.

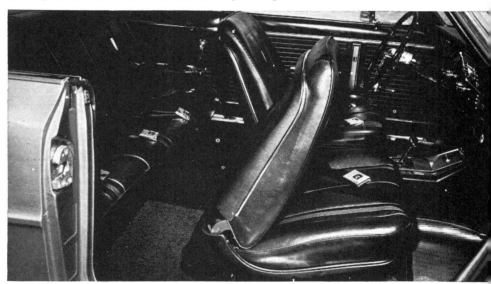
Rear bench seat is roomy enough for three, but preferably two for any duration. Front buckets are pleasure to sit in for any period of time and provide "just right" support.

Padding on top of dash is standard, as is non-glare paint on exposed metal. Inside door handle serves as lock mechanism when pushed down. Key must be used from outside.

CHEVY II NOVA 327

caught it before it was released to us. Because of this, and the lack of a hose within a day's drive of where we were, acceleration tests for the Chevy II were over. The damage was temporarily mended with some friction tape.

Later, we looked at a number of other 327 Chevy IIs and they all had the correct hose, so we believe that ours was "just one of those things." We also believe if we had been able to get in another couple of passes, the car would have run the quarter in less than 16 seconds, and the rest of the acceleration figures would have improved accordingly.

As you can see by the performance charts, the Nova didn't perform badly — considering. And it didn't stop badly, either. From 60 mph, some wheel hop is evident from the rear end. This could be a feature of the single-leaf spring back there. It does stop straight from 60 as well as 30 mph, and there is no reduction of pedal feel during hard stops. They're drum brakes, of course, but with a car this light, they will provide adequate braking under almost any circumstance a driver may encounter.

We found the Chevy II had rather minimum, minimum ground clearance. Trying to go down a ramp that we'd not had trouble with on any other car, the head pipe of the exhaust system scraped. We measured it on level ground, and it was a scant 5.5 inches. Some other cars also have a 5.5-inch clearance, but it is usually at a place under the rear or front end. When it is in the middle of the car, trouble of the kind we described is a sure thing. The front and rear overhang isn't great, so bottoming out or scraping didn't happen there.

The Chevy II Nova, with its 327, provides much that we have been complaining other cars lacked. The overhead dome light is bright enough to do your homework by, arm rests are at all four corners and they do rest your arm when used, the ash tray is convenient to the driver and is sufficiently deep to double as a litter bag, and the seats are comfortable enough to ride in all day long without becoming numb in certain areas. Parallel parking isn't a chore as the driver can see the tops of the front fenders clearly, and there is no great side overhang to the body.

The Chevy II 275-hp Nova is a car of many virtues. It's practical and safe enough for a family car, sporty and peppy enough for any young man on date night, and economical enough for about everyone's budget. —*Steve Kelly*

Side profile reveals smooth lines and absence of polish-hungry chrome trim. All Chevy IIs with V-8s have 14-inch wheels.

Front bumper is of same massive size as rear, which should go far in protecting the extruded aluminum grille from damage.

PHOTOS BY DARRYL NORENBERG

It's a tight fit for engine compartment to accept a 327. Spark plugs, in most instances, have to be replaced from underneath.

Coil springs in front, single leaf springs in back give unit-body construction good ride. The resonator rides near fender.

CHEVY II NOVA 327

MT ROAD TEST

SPECIFICATIONS FROM MANUFACTURER
ENGINE IN TEST CAR: Ohv V-8
  Bore and stroke: 4.001 x 3.250 ins.
  Displacement: 327 cu. ins.
  Advertised horsepower: 275 @ 4800 rpm
  Max. torque: 355 lbs.-ft. @ 3200 rpm
  Compression ratio: 10.5:1
  Carburetion: 1 4-bbl.
TRANSMISSION TYPE & FINAL DRIVE RATIO: Powerglide (torque converter with planetary gears). Floor-mounted shift lever. 3.08:1 rear-axle ratio.
SUSPENSION: Independent front with coil spring. Lower control arm strut-supported. Hotchkiss rear suspension with 2 single-leaf rear springs. Direct-acting tubular shocks at each wheel.
STEERING: Semi-reversible, recirculating ball nut; linkage-type power assist.
  Turning diameter: 38.4 ft., curb to curb
  Turns lock to lock: 4.5
WHEELS: Short-spoke disc steel
TIRES: 6.95 x 14 4-ply-rated rayon tubeless
BRAKES: Duo-servo hydraulic; self-adjusting
  Diameter of drum: front, 9.5 ins.; rear, 9.5 ins.

SERVICE:
  Type of fuel recommended: Premium
  Fuel capacity: 16 gals.
  Oil capacity: 4 qts.; with filter, 5 qts.
  Shortest lubrication interval: 6000 mi., or 60 days
  Oil- and filter-change interval: 6000 mi., or 60 days
BODY & FRAME: Unitized construction, with front end rigidly bolted to body proper. Frame members incorporated.
  Wheelbase: 110.0 ins.
  Track: front, 56.8 ins.; rear, 56.3 ins.
  Overall: length, 183.0 ins.; width, 71.3 ins.; height, 53.8 ins.
  Min. ground clearance: 5.5 ins.
  Usable trunk capacity: NA
  Curb weight: 3040 lbs.
NA — Information not available at presstime

PERFORMANCE
ACCELERATION (2 aboard)
  0-30 mph .... 3.4 secs.
  0-50 mph .... 6.5 secs.
  0-60 mph .... 8.6 secs.
  0-75 mph .... 12.9 secs.
TIME & DISTANCE TO ATTAIN PASSING SPEEDS
  40-60 mph .... 4.1 secs., 299.3 ft.
  50-70 mph .... 4.8 secs., 422.4 ft.
STANDING-START QUARTER-MILE: 16.4 secs. and 85.87 mph
BEST SPEEDS IN GEARS @ SHIFT POINTS
  1st .... 58 mph @ 4600 rpm
  2nd .... (not maximum) 78 mph @ 3400 rpm
MPH PER 1000 RPM: 22.9
SPEEDOMETER ERROR AT 60 MPH: 4% fast.
STOPPING DISTANCES: from 30 mph, 30.5 ft.; from 60 mph 152.5 ft.

ACCESSORY PRICE LIST
Engine options: 140-hp 6-cylinder .... $ 26.35
            to 350-hp V-8 .... 198.05
*Automatic transmission 172.92
 4-speed transmission 184.35
 Overdrive .... —
 Limited-slip differential 36.90
 Heavy-duty suspension 4.75
*Whitewall tires 28.03
 Disc brakes .... —
*Power brakes 42.15
*Power steering 84.30
 Power windows .... —
 Power seat .... —
*Radio AM 57.40
 Radio AM/FM .... —
 Air conditioning 310.70
*Tinted glass 26.35
*Bucket seats (Nova SS only) .... std.
 Adjustable steering wheel .... —
*Clock (less installation) 21.60
 Tachometer (less installation) 39.95
 Automatic headlight dimmer .... —
 Automatic speed regulator .... —
 Vinyl roof cover 73.75
 Head rests (Bucket-seat type) 52.70
*On test car
Dash (—) — not offered

MANUFACTURER'S SUGGESTED LIST PRICE: $2517 (incl. taxes, safety equip't & PCV device)
PRICE OF CAR TESTED: $3165.15 (incl. excise tax, delivery & get-ready charges, but not local tax & license)
MANUFACTURER'S WARRANTY: 24,000 miles and/or 24 months

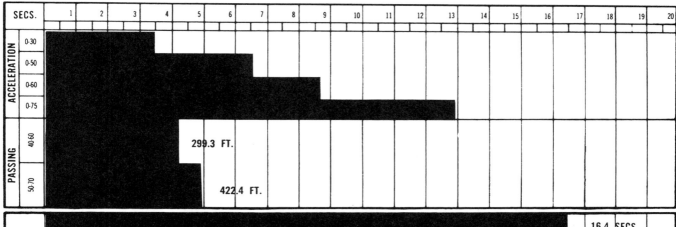

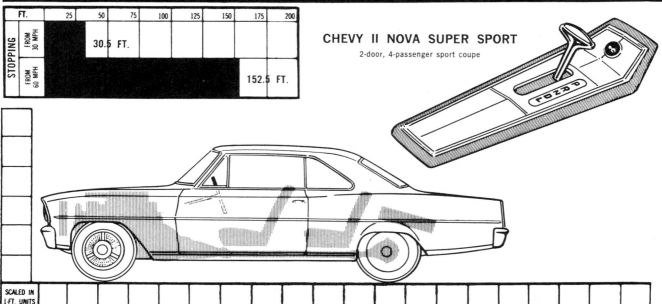

CHEVY II NOVA SUPER SPORT
2-door, 4-passenger sport coupe

MOTOR TREND / JULY 1966

1968 report

CHEVY II

Longer wheelbase supports a new body in the semi-fastback, short deck, long hood concept for '68.

With longer wheelbase (111 inches) and wider tread, the new fastback Chevy II offers exceptional handling.

Chevy II still sports single headlights. Nova SS has black accented grille with SS emblem dominant.

Hardtop has been dropped for '68, but new lines on coupes keep it from being missed.

Styling changes for the Chevy II are based on the short deck-long hood concept sweeping Detroit this year. The design, which incorporates a semi-fastback roof line, gives the Chevy II a new, light appearance. Based on a longer wheelbase of 111 inches, the car is lower and around six inches longer than last year's model.

For the first time, Chevy II will feature curved side windows, foot-operated parking brake and keyless locking on all doors. This is to be part of a general upgrading of the Chevy II line.

Top of the line is the Nova SS coupe since the hardtop model has been dropped for '68. The Nova SS is available with the 295-horsepower version of the 350-cubic-inch V-8 only. Available with three-speed, heavy duty three-speed, four-speed or Powerglide transmissions, and heavy duty suspension, it's a ball to drive. Power output is solid and handling is exceptional. Part of this is due, on standard models, to a larger tire size, 7.35 x 14. Nova SS gets the 14-inch wide oval tires.

The Nova is available in either coupe or four-door sedan models only. Bucket seats are optional on the coupe and a host of additional options inside and out allow the owner to custom tailor the Chevy II to his taste. Six cylinder engines are standard on the Nova and top of the line is the 275-horsepower 327 V-8. Standard V-8 is Chevy's new 307-cubic-inch engine with 200 horsepower. It's a regular fuel consumer.

New fastback roof on coupes limits vision to rear but styling is greatly improved.

The 325-hp 327 Chevy II is a performer in its own right—and you can "optionalize" a II from sweet to sassy • by Steve Kelly

CHEVY'S POP COMPACT

Heralded as an "economy" super-car in '66, when out-fitted with a 327 cubic inch, 350-hp engine, the compact Chevy II lost good billing in '67 when that engine was dropped from the option list. Chevrolet bounced back firmly into the mini-super area in '68 with their reshaped Chevy II and a better line-up of engine combos for the whole series. When first introduced last fall, a 350 cubic inch, 295-hp V8 was the top performer, but it at least hinted that the car hadn't been forgotten in the performance planning area. Not long afterward, Chevy's 325-hp, 327 cubic inch V8 was thrown onto the power-plant stack, and promise of what it could do led us in search of a "II" so equipped. Announced after our test was completed is a 396 V8 option. A report on it is in this issue's " '68½ Chevys."

Our test car was a Nova two-door coupe with a good amount of deluxe trimmings. Incidentally, only a two-door coupe with pillar and a conventional four-door sedan are offered in the '68 Chevy II.

The car, when first received, was in a state of almost total "un-tune," and we had to do some catch-up work on the engine before it would even operate smoothly. It took a thousand miles or so for it to loosen up, and then we started lead-footing it some to get an idea of just how the car might perform.

First trips through the quarter-mile were disappointing. The best time we achieved — in un-touched form, but everything set at factory specs — was 15.7 seconds, and a speed of 89 mph. Grim, but we knew it could be bettered. We made an appointment with Bill Thomas to work in a few of his tricks on the Nova, and just drove the car in town for a day or so until the appointed hour.

Plain and simple driving is extremely pleasurable, with near-perfect seat altitude and vision far less obstructed than in many current fastbacks. The high seat position, though, will cause some head-bumping by average-sized humans, until they remember to bow their heads a bit getting in and out. The upper door line is rather low, though not a vision hindrance while driving.

Steering wheel placement is too high for comfortable hand grip and arm position. It would be much worse were it not for the seat altitude, and even then it's in need of lowering. An adjustable steering column is the answer, but this isn't offered for Chevy II's. As it turned out, we grew accustomed to the wheel, but we're thankful our coupe had the higher-placed bucket seats and not a lower bench-type.

Overall comfort is great. We had occasion to put many miles on the car in single stretches, and not once grew sore on our backsides from lack of cushioning.

Pedal position could be improved. Elevation of the clutch pad to the floor is high, and conscious movement goes into the act of pulling your left foot from the floor to the clutch. Our power brake system had its pedal placed lower than the clutch, which is good, but prohibited any "heel-and-toeing" from gas to brake, since the accelerator is lower still. Also, the metal banding (decorative) on the clutch and brake proved treacherous with wet shoes. One morning we jumped into the car out of the rain, pushed the clutch pedal down and had it pop back up and tear out a good chunk of skin over our shins, after our foot slipped right off the pad. All that rubber down there doesn't do much good if metal contacts shoe leather. Good way to sabotage your enemies, though.

Instrument clustering is good only for speedometer, clock and tach — if the optional gauges are ordered. If the optional units are part of the package, then there's good cause for a few gripes. The tach is in the left corner of the main dial face, directly in front of the driver where it should be. But the fuel, oil, amps and temperature gauges are housed on top of the also-optional floor-mounted console. And with a black interior, chances of your seeing readings in a hurry after looking out on brightly lit scenes are nil. Most of the time, even at night, you must take second looks at the square-faced instruments to register what they're saying. Otherwise, the actual dashboard face is clean and uncluttered. The radio is housed low, right in front of the driver.

About the only other serious gripe we can make on the well-designed car is the shift linkage. After missing shifts a few times (few?), we'd almost decided to write a linkage story and include the Chevy II road test. Short of churning butter, there's no more "mysterious" feeling than that you encounter when trying to smoothly — and quickly — slide the four-speed Muncie shifter from gear to gear. The main shift unit is mounted remote from the transmission, and when the engine torques, the shift-actuation stays still. This can set up a bind between shift handle and connecting arms, changing geometry to the detriment of smooth operation. The throws are long, and 1-2 are widely separated from 3-4, with a less-than-good "gate." We'd like to see Chevrolet offer a quality extra-cost unit such as Olds and Buick now have, but short of that, all we can advise is to save some extra bucks for a good unit to go in after the car leaves the showroom.

The silver coupe only stayed at Thomas's shop a few day and Bill's work was relegate to new plugs, slightly larger je ting, a distributor rework wit a curve set to bring in tot advance of 39 degrees at abo 3000 rpm, hydraulic lifter a justment and some smoothin out of the gearshift linkage which was greatly appreciate

We revisited Irwindale Rac way for another try, leaving th car in exactly the same trim had been before the tune. O first pass was a 15.39, and w lowered this to a 15.30 e.t. the next run. We pulled the a cleaner off, though this didr cut more than .05-second, an then disconnected the sm control pump and power stee ing. Doing this brought a le of 15.10. We had 40 pounds a pressure in the front tires, a 32 in back. We dropped t rear tire pressure to 28 befo installing slicks, but the dr slowed us. At this point, o slowest mph reading had be 91.74, and our fastest — corded with the 15.10 e.t. was 95.33.

Our next experiment dea with putting a set of 27-in Casler Stock slicks on Crag SS wheels. 'Course by this tim we already had Cragars up fro and the car looked 100 perce better for it. Getting the slic on is no problem, but the offs Cragar wheels caused the si walls of the 7-inch cap slicks rub the inner fender wells. N the wheel well lip, but the insi of the fender. The only soluti for sustained running of w tires and offset wheels on t car would be to radius the f fender well with the tire. B our usage was for testing on and as long as we didn't ha any weight in the trunk or pa sengers inside, the tires did rub. So we left the sheet me intact.

First pass down the quar with the slicks netted us

HOT ROD MAGAZINE

LEFT — Under-hood view is complicated by smog plumbing, but access is still good and chrome trim sparks appearance. CENTER — Even emission controls can't keep a good torquing small-block from smokin' 7-inch Caslers. Cragar-mounted slicks wrinkled easily even with high air pressure. BOTTOM — While stylists did good esthetic job on dash, practicality was slighted.

photography: Eric Rickman

VEHICLE
Chevy II Nova

PRICE
Base $2367.00
As tested $3772.90

ENGINE
Type . . . . OHV V8
Cylinders 8
Bore & stroke . . . . . . 4.001 x 3.25 in.
Displacement . . . . 327 cu. in.
Compression ratio 11.0:1
Horsepower . . . . 325 @ 5600 rpm
Torque . . . . 355 lbs.-ft. @ 3600 rpm
Valves: Intake . . . . 2.020 in. dia.
Exhaust . . . . 1.600 in. dia.
Camshaft:
    Lift . . . . . .4471 in., intake and exhaust
    Duration . . 306°, intake and exhaust
    Overlap . . . . 76°
Tappets . . . . Hydraulic
Carburetion . . . . Rochester 4-bbl
Exhaust system . . . . Dual

TRANSMISSION
Type . . . . . 4-speed manual, all-synchro in forward gears. Floor mtd. shifter
Ratios: 1st 2.54:1
    2nd .1.88:1
    3rd 1.46:1
    4th 1:1

DIFFERENTIAL
Type . . . . Semi-floating, overhung pinion gear. Positraction
Ring gear diameter . . . . 8.875 in.
Ratio 3.55:1

BRAKES
Type . . . . Front disc/rear drum with power assist
Dimensions: Front disc . . . . 11.0 in.
    Rear drum . . . . 9.5 in.
Swept area . . . . .332.4 sq. in.
Effective area . . . . .114.0 sq. in.
Percent brake effectiveness
    front 58.5%

SUSPENSION
Front . . . . Independent, single lateral arm type with coil spring
Rear . . . . .Salisbury rear axle with multi-leaf, semi-elliptic springs
Shocks . . . . . . Direct acting, tube type
Stabilizer . . . . Front only, link type, .687 in. dia.
Tires . . . . . 7.35x14 conventional pattern
Wheel rim width . . . . .5 in.
Steering:
    Type . . . Semi-reversible recirculating ball nut, linkage-type power assist
    Gear ratio 17.5:1
    Overall ratio 20.7:1
    Turning circle . . . 38 ft., curb to curb
    Wheel diameter . . . . 16.5 ins.
    Wheel turns lock to lock 3.5

PERFORMANCE
Standing start quarter-mile (best)
. . . . 14.60 sec., 95.33 mph

DIMENSIONS
Wheelbase . . . . .111.0 in.
Front track . . . . 59.0 in.
Rear track . . . . 58.9 in.
Overall height . . . . 54.1 in.
Overall width . . . . 72.4 in.
Overall length . . . . 189.4 in.
Curb weight . . . . 3410 lbs.
Body construction . . . comb. body/frame with forward frame.
Crankcase capacity w/filter. . . . . . 5 qt.
Cooling system . . . . .16 qt.
Fuel tank . . . . .18 gal.

CHEVY'S POP COMPACT

14.92, but a slower speed of 93.45 mph. The Caslers were taller than the stock skins, so we had expected a slower speed. We'd been coming off the line at about 2000 rpm with the stock tires, but were able to up this to 4500 rpm with slicks. We varied air pressure, but finally held firm with 26 pounds in the Caslers. Our times varied considerably on the run sheet, thanks altogether to missed shifts. We're not in the habit of hanging-up between gears, and got more than irritated at the linkage each time this occurred.

Our best run with slicks, belts off and air cleaner removed, was 14.60 seconds elapsed time and a best speed of 94.33 mph. Our speed dropped exactly one mile per hour, but e.t.'s got better by exactly one-half second.

High-placed bumper reveals sturdy underpinnings from low angle, and does help usher cool air to engine. Slots in bumper help here too. Front end reflects trim styling theme evident on entire car.

A little shorter tire would've helped, but not many people offer one. A better bet, for drag racing, would be a lower gear. Our coupe came with a 3.55:1 Positraction axle, and special orders can be made on assembly-line cars to get them equipped with axles running to 4.88:1.

This car weighed 3410 pounds in street trim with a full tank of gas. It carries a factory shipping weight of about 3020 pounds, so it's easy to see how much can be trimmed and still keep it legal. The bucket seats, power disc brakes, instrumentation, radio, power steering and the host of other goodies this had, jacked up the weight considerably, but we wouldn't want to use it long in traffic without these touches.

Drag racing's not everyone's bag, and there's a lot more to this car than just good straight-line potential. It scores well in the handling department, exhibiting very little lean, and oversteer only on the sharpest of turns at a brisk pace. We admire its agility and can easily see a vast improvement over earlier Chevy II's. Four-speed-equipped 327's, and all 350's and 396's have multiple-leaf springs instead of the once-famous single-leaf design initiated in '62. We didn't encounter a trace of wheel hop on acceleration, even with the slicks in place, or upon deceleration. That's a lot more than could be said for the V8-powered Chevy II of yesteryear. In-town driving presents no drastic disadvantage with the Muncie shifter, as it will downshift without kicking it. We remember sampling a Firebird Sprint once with a standard Muncie unit that virtually refused to move any faster than an inch per second.

Braking is excellent. The power front discs halted us from our near-100-mph drag strip speeds run after run without diminishing, and allowed us to grab the second return road opening every time. They are rather quick to react, but don't have the instant-vise action of many. Stops from 60 and 70 mph revealed lock-up only as the car passed 30 mph, and even then, front wheel maneuverability was maintained.

Rear seat passengers might bellyache if they're nearly six feet tall. Head room is marginal, but the knee room is pretty slim. Coupled to this is the hard-backed bucket seats that press against rear occupants' knees, instead of vice versa.

Once inside, we got the feeling of being in a much larger car. The relatively high overall height and well-placed seats allow drivers to view the complete length of the hood. Also, they get a chance to see things a bit better and sit a bit higher than in other current domestic-built autos. This adds to agility, both physical and mental. If you can see it sooner, you'll react sooner.

Parking isn't that much of a chore, even though it has a semi-fastback roofline. Reason for its being better — than a Chevelle for instance — is the shorter rear deck and the closer position of the driver to the back glass.

While others have vacated the compact field, the Chevy II remains solid in this marketplace, though it has grown some since its introduction. 1968 is the first completely new design, but all the good things established earlier remain. Like size, and comfort, and luggage space to handle a young family. Economy isn't what a 325-hp engine calls for, but we were surprised. Initial readings were down to about 12 miles per gallon, but we reached a high of 17.4 after the tune job. The hydraulic-lifter engine operates smoothly without excessive noise. There's still a hint of radical cam timing emanating from the muffled and resonated exhaust, so all is not lost in quietness. The compact price is still there too, with a base of around $2,300 showing at the top of the window sticker. That's a four-banger (yep, they're still making them) base, but at least it's a nice low place for beginning.

We dropped by the Southern California Automobile Club one afternoon for a speedometer check, where we found a 3-mph pessimistic error, but also got hooked up to a smog emission control tester in the process. The potent Chevy came out excellent here, registering 135 parts per million of unburned gasoline, against a standard of 275 p.p.m. Now, if we can only get down to 0 on the unburned gas count, we'll really kick out the horsepower.

Going quicker than 14.6 with a 325-hp Chevy II should be simple. Building in a better distributor lead will help, so that it's all in by 2000-2200 rpm. And a set of headers will probably show the most improvement. Running "corked-up" as we did, we forced all those unburned gases through a dual exhaust and cross-mounted single muffler. Then a lower-ratio rear axle, proper head blueprinting, some suspension adjustments (please don't raise 'em up any more), and of course a first-class shifter should have a similar machine dipping toward the very low 13's, yet still streetworthy.

If you don't feel this is the way to go, take it from us; there're a few kicks available just driving around with the radio on. Get better mileage that way too. ∎∎

HOT ROD MAGAZINE

SOMEONE on the Car Life staff asked, "What kind of car is that Chevy II—is it a Supercar, a Ponycar, or what?" And, after much head scratching, we never did come up with an answer. Our test car was a Chevy II Nova SS coupe, with 327-cid/325-bhp V-8 engine, four-speed manual transmission, power disc brakes, and 3.55:1 Positraction rear axle. Obviously, a sportingly equipped automobile.

In a sense, the Chevy II reminded us of some of the better imported sports sedans, like the Volvo 144, BMW 1600-1800, Rover 2000, and others. Whatever the category, the Chevy II was an extremely satisfying automobile, a car with enough interior room and luggage space for the family, but small enough to give the driver a true sense of command. Before imported car fans begin firing letters extolling the virtues of their particular example of sports sedan, they should remember we said "reminded," not "matched" or "duplicated."

A brief test drive in an early Chevy II at the GM Proving Grounds last summer convinced us that this was a fine handling car. So, we selected a car with the options that we felt best complemented these inherent sporting characteristics. During the course of this road test, the initial impressions of good handling held true, and we

THE CHEVY II THAT OPTIONS BUILT

In many ways it reminded us of a European sporting sedan

FULL-THROTTLE CORNERING of handling optioned Chevy II produced oversteer, but reverse steering lock kept car stable.

TRAILING THROTTLE cornering brought out predictable understeer, easily controlled on smooth test track, but less secure on rough roads. Body lean felt extreme, but Chevy II leaned this far, then leaned no further.

CHEVY II

SUSPENSION OPTIONS made Chevy II stable at high speed. The Nova SS package brings stiff springs, firm shock absorbers, oversize tires and a harsh ride.

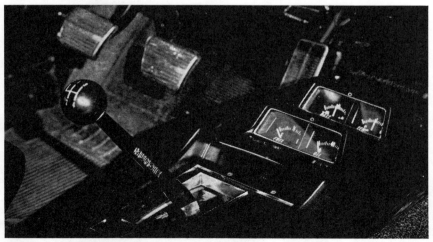

GEAR SELECTION was awkward, with first and third gears impossible to reach without leaning forward. Console gauges were $95 option and were difficult to see.

found ourselves spending an inordinate amount of time racing around the road race course at Orange County Raceway—and enjoying every lap.

The Chevy II was another car that pointed out the fallacy of buying a 4000-lb., full-size domestic sedan just to get a reasonable amount of passenger room and luggage accommodation. A family of five—two adults and three children found the Chevy II a perfectly satisfactory family vehicle, whether for around-town errands or extended highway cruising. The test car was a two-door sedan, though families with large children may find the four-door model more suitable. Head room in Chevy II two-door versions is a bit scant in the rear.

The wide, deep trunk of the Chevy II was truly an unexpected bonus. The only flaw was the vulnerability of the rear fender sheetmetal at the sides of the trunk compartment. With no trunk inner panels, cargo sliding about in the trunk inflicted inside-out dents in the flanks of the body shell. It would be wise to install some sort of barriers in the outer edges of the trunk floor. This same comment applies to almost all domestic cars—the Chevy II was simply a prime example.

Driving the Chevy II again emphasized its comparability to foreign sedans. The driver sits quite high, and rather erect. Steering wheel location is high, with the wheel angle sharply forward at the top. We found ourselves thinking back to the early 1950s, when "chair-height" seating was an advertising point for Chrysler Corp. A high, erect driving position may not give one the feeling of "sports car environment," but it enhances both comfort and visibility.

Control location in the Chevy II was judged acceptable, but not outstanding. The gearshift is too far forward for engagement of first and third gears without leaning forward. Power brakes gave a low brake pedal, but the clutch pedal was much too high. Since the accelerator pedal was even lower than the brake, the driver needed a left leg that was a few inches shorter than the right. Invariably, a driver selected a compromise location for the seat in town, then moved the seat rearward for highway cruising where clutch pedal operation and gearshift reach were not so critical. (The Chevy II made it painfully obvious that Detroit auto executives never drive cars with manual transmissions, but confine themselves to "more civilized" automatics.)

The actual seat designs, both front (bucket) and rear (bench), were fair. Comfort was acceptable, with seatback rake angle a bit too vertical for sustained cruising. Lateral support of the front buckets was almost nonexistent for average-size drivers.

The high Chevy II driving position at first gave an impression of tippiness and instability. This soon gave way to one of complete confidence. The test car was fitted with the Nova SS package, which includes heavy-duty springs and shock absorbers, and oversize tires. This is a combination we highly recommend. High-speed stability of the test car was excellent, and handling was near perfect.

If a curve was entered at a fairly low speed, and moderate power applied, the Chevy II went into a strong understeering plow. If the same curve was entered near the maximum permissible cornering speed, the Chevy II hooked into the bend and set itself in a nearly neutral attitude. Tossing the car into a bend even faster, and countering with full throttle application, moved the Chevy II's tail out like "ole A.J." on a dirt oval. The resulting power-oversteer slide could be held, increased or decreased with the throttle, and a driver would have had to try awfully hard to spin the car completely. We never did.

Steering response, overall agility and cornering attitude control of the Chevy II were, again, on a par with some of the best imported sports sedans, and a cut above domestic intermediates. The Chevy II's lack of unnecessary bulk and overhang made it nimble and maneuverable, a delight to thread through heavy traffic or storm through a winding mountain pass.

SMALL ENGINE fits easily into big compartment. Highway performance was more than adequate, but acceleration times were disappointing.

SPACIOUS TRUNK was a bonus, with room for plenty of luggage, despite location of spare tire. However, rear panels are easily damaged by loose cargo.

While wringing out the Chevy II on the tricky, but very fast, road course at Orange County International Raceway, a pair of English motorcycling journalists appeared. They watched Engineering Editor Jon McKibben pushing the Chevy II through the turns, and forthwith requested a ride with him. Climbing in and buckling up, they charged off. A lap later, they stepped from the car in the pit area, pale and shaking, but impressed with the handling of this "American cooking sedan." In fact, they couldn't believe that this was the way the car came from the assembly line. Cars simply didn't handle that well without extensive suspension system rework. But the Chevy II did handle, at least as well as the Britishers' beloved Jaguar "saloons."

The place where our imported sports sedan analogy fell apart was in ride and handling over moderately rough roads. Where a car like the BMW sedan has a thoroughly developed fully independent suspension system to keep all four wheels in firm contact with irregular pavement, the

CHEVY II

Chevy II has unsophisticated and too-stiff suspension that allows the rear wheels to spend too much time in the air instead of on the road. Also, the stiff springs used to achieve stable cornering yield a back-slapping, harsh ride. Finally, an excessive amount of chassis/body shake was noted, the front sub-frame and suspension components apparently moving about under the unitized shell. Nor did the surface have to resemble a bomb-razed battlefield. One of our favorite fast back road bends has a ridge running across the pavement, near the normal apex of the turn. This ridge, approximately an inch high, caused the Chevy II to lift free of the road and return some two or three feet outside the original cornering line.

Rear axle control was deficient for all-out acceleration runs. Chevrolet claims its staggered shock absorber arrangement obviates the need for any sort of trailing arms for axle hop control. This proved incorrect on the test car, as every hard shift was accompanied by multiple chirps from the bouncing rear tires and an unpleasant lurching from the rear suspension. The rear axle did not hop, though, on powering out of turns. Or, in fact, under any circumstances likely to be encountered on the street.

The test car was equipped with the excellent Muncie four-speed manual transmission. Unfortunately, the shift linkage was the same kind that has been fitted to other Chevrolet products (except Corvettes) for the past two years. It is amazing that the same transmission that shifts so effortlessly and smoothly in Pontiacs, with Hurst linkage, can shift so badly in Chevrolets using their linkage with that sliding plastic plate in the console.

Another disappointment was found in the performance of the 325-bhp engine. This engine should be the hottest thing available in a Chevy II, offering more peak bhp than the 350-cid/295-bhp option. In fact, the 325-bhp Chevy II option is the same engine as the 350-bhp Corvette option (L79). Paper horsepower again.

Top drag racers around the U.S. have been turning consistent 12-sec. quarter-miles with stripped Chevy IIs with this engine, so the potential is obvious. That our test car could only manage 16.47 sec. elapsed times indicates several things: full emission control equipment creates very lean mixtures, resulting in significant loss in torque; relatively narrow-section tires (7.35-14 B. F. Goodrich Silvertown) cause severe traction loss, particularly with the light-tailed Chevy II; balky gearshift linkage forced slower-than-normal gear changes; standard 3.55:1 axle ratio prevented the relatively low torque engine from accelerating the 3770-lb. (test weight) car at a rapid rate; and standard exhaust coupled to the low-mileage engine restricted output at high rpm, giving a strained feeling above 5000 rpm.

Even though dragstrip performance wouldn't scare Supercar owners, the Chevy II had ample power for passing, and was fast enough to be fun to drive. Well, it should have been. The test car had $1275 worth of options added to the $2367 base price. Of these, about $500 worth of performance options were included. The remainder of the list was composed of comfort and convenience accessories, with a $95 special instrument package incorporating oil pressure, water temperature and fuel gauges and an ammeter. These were located down on the console where they could be given a token glance occasionally. A miniature tachometer was housed in the normal fuel gauge slot in the instrument panel. Because of its size, the tach was useful only for approximating engine speed. ∎

1968 CHEVROLET
CHEVY II NOVA SPORT COUPE

DIMENSIONS
Wheelbase, in. 111.0
Track, f/r, in. 59.0/58.9
Overall length, in. 187.7
 width. 70.5
 height. 54.1
Front seat hip room, in. .... 23.2 x 2
 shoulder room. 56.9
 head room. 41.6
 pedal-seatback, max. 40.5
Rear seat hip room, in. 50.3
 shoulder room. 55.0
 leg room. 32.6
 head room. 36.6
Door opening width, in. 37.4
Trunk liftover height, in. ... 30.8

PRICES
List, FOB factory $2367
Equipped as tested $3642
Options included: 325-bhp/327-cid V-8, four-speed transmission, strato-bucket seats, power steering and disc brakes, special instrumentation, am radio, HD radiator, limited slip differential.

CAPACITIES
No. of passengers 4
Luggage space, cu. ft. .... n.a.
Fuel tank, gal. 18
Crankcase, qt. 4
Transmission/dif., pt. 3/3.5
Radiator coolant, qt. 17

CHASSIS/SUSPENSION
Frame type: Unitized, front sub.
Front suspension type: Independent by s.l.a., coil springs and concentric shock absorbers.
 ride rate at wheel, lb./in. ..... n.a.
 antiroll bar dia., in. 0.687
Rear suspension type: Hotchkiss live axle, multileaf springs, telescopic shock absorbers.
 ride rate at wheel, lb./in. ..... n.a
Steering system: Integral assist recirculating ball gear, parallelogram linkage behind front wheels.
 overall ratio 20.6:1
 turns, lock to lock 3.6
 turning circle, ft. curb-curb ... n.a.
Curb weight, lb. 3400
Test weight 3770
 distribution (driver),
 % f/r 53.4/47.6

BRAKES
Type: Ventilated disc front cast iron duo-servo drum rear, proportioning valve.
Front rotor, dia. x
 width, in. .... 11.0 x 2.21
Rear drum, dia. x width .... 9.5 x 2.21
 total swept area, sq. in. 332.4
Power assist: Integral vacuum.
 line psi at 100 lb. pedal ...... n.a

WHEELS/TIRES
Wheel rim size .... 14 x 5J
 optional size .... none
 bolt no./circle dia. in. 5/4.75
Tires: B.F. Goodrich Silvertown 660.
 size 7.35-14
 normal inflation, psi f/r 28/28
Capacity @ p.s.i. .... 5040 @ 28

ENGINE
Type, no. of cyl. .... ohv 90° V-8
Bore x stroke, in. .... 4.001 x 3.25
Displacement, cu. in. 327
Compression ratio 11.0:1
Fuel required .... premium
Rated bhp @ rpm .... 325 @ 5600
 equivalent mph 116
Rated torque @ rpm ...... 355 @ 3600
 equivalent mph 74
Carburetion: Rochester 1x4.
 throttle dia., pri./sec. 1.38/1.72
Valve train: Hydraulic lifters, pushrods and overhead rocker arms.
 cam timing
 deg., int./exh. 40-86/88-38
 duration, int./exh. 306/306
Exhaust system: Dual, two reverse flow mufflers.
 pipe dia., exh./tail. 2.50/2.25
Normal oil press. @ rpm . 60 @ 2000
Electrical supply, V./amp. 12/37
Battery, plates/amp. hr. 66/61

DRIVE TRAIN
Clutch type: Single dry disc, semi-centrifugal.
 dia., in. 10.34
Transmission type: Four-speed, all synchromesh.
Gear ratio 4th (1.00:1) overall. 3.55:1
 3rd (1.46:1) 5.18:1
 2nd (1.88:1) 6.67:1
 1st (2.52:1) 8.95:1
Shift lever location: Floor.
Differential type: Hypoid, limited slip.
 axle ratio 3.55:1

CAR LIFE

Chevy II SS 396

Quick in a straight line, but ill-suited to anything else. That sums up the Chevy II SS396, new addition to the Chevy II line.

This is basically a Chevy II Nova SS coupe, with the 396-cid Chevelle and Camaro engine. The car we recently drove was equipped with a 350-bhp version, but presumably the 325- and 375-bhp engines offered in Camaro and Chevelle models will also be available in the Chevy II.

Obviously, the Chevy II SS396 is a fast accelerating vehicle. However, it is not the fastest quarter-miler on earth. And this takes away the reason for being for the car. The SS396 handles as one might expect—the extreme forward weight bias causes severe understeer and plowing. Enough torque is available to break the rear end loose, but the car is so poorly balanced, and standard front tires so woefully overstressed in hard cornering, that motoring rapidly through a series of turns is more chore than fun.

The other ill effect of the 396-cid engine is very poor brake proportioning. Attempts to pull high-deceleration stops met with rear wheel lockup and drastic loss of directional control. Apparently, brake proportioning for the SS396 is the same as for cars with much lighter, small-block engines. ∎

CAR LIFE ROAD TEST

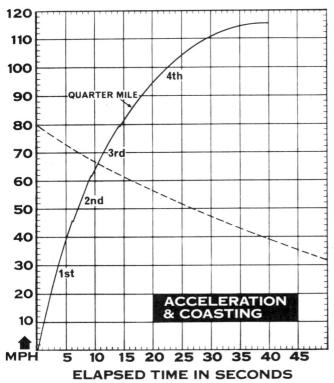

CALCULATED DATA

Lb./bhp (test weight) 11.6
Cu. ft./ton mile 145.6
Mph/1000 rpm (high gear) 20.7
Engine revs/mile (60 mph) 2900
Piston travel, ft./mile 1570
CAR LIFE wear index 45.5
Frontal area, sq. ft. 21.2
NHRA-AHRA class .... C/S-n.a.

SPEEDOMETER ERROR

30 mph, actual 30.7
40 mph 41.7
50 mph 52.0
60 mph 62.5
70 mph 74.4
80 mph 82.6
90 mph 93.8

MAINTENANCE

Engine oil, miles/days 6000/120
oil filter, miles/days 6000/120
Chassis lubrication, miles 6000
Antismog servicing, type/miles ..
clean air pump system/12,000, replace PCV valve/12,000
Air cleaner, miles replace/24,000
Spark plugs: AC44.
gap, (in.) 0.033
Basic timing, deg./rpm ... 4BTC/700
max. cent. adv., deg./rpm.30/4700
max. vac. adv.,
deg./in. Hg. 15/15.5
Ignition point gap, in. 0.019
cam dwell angle, deg. 30
arm tension, oz. 21
Tappet clearance, int./exh. 0/0
Fuel pressure at idle, psi 6.0
Radiator cap relief press., psi 15

PERFORMANCE

Top speed (5600), mph 116
Test shift points (rpm) @ mph
3rd to 4th (5600) 79
2nd to 3rd (5600) 62
1st to 2nd (5600) 46

ACCELERATION

0-30 mph, sec 3.7
0-40 mph 5.1
0-50 mph 6.8
0-60 mph 8.7
0-70 mph 11.4
0-80 mph 14.5
0-90 mph 18.5
0-100 mph 22.5
Standing ¼-mile, sec 16.47
speed at end, mph 86.0
Passing, 30-70 mph, sec 7.7

BRAKING

Max. deceleration rate from 80 mph ft./sec² 24
No. of stops from 80 mph (60-sec. intervals) before 20% loss in deceleration rate .... 8-no loss
Control loss? Moderate.
Overall brake performance ...... fair

FUEL CONSUMPTION

Test conditions, mpg 12.8
Normal cond., mpg 12-15
Cruising range, miles 190-250

DRAG FACTOR

Total drag @ 60 mph, lb .... n.a.

MAY 1968

Foul weather awaited brave group after departing amidst Angels Camp's morning rush.

CHEVY II VS VW

by Steve Kelly

"How do you think a Chevy II would stack up against the new, fuel-injected VW 1600 in economy?"

"Don't know," we replied to the questioner on the other end of the line, "but why do you ask?"

"Well," came the reply from Dick Wilmshurst, president of 49er Chevrolet, Angels Camp, Calif., "I'd like to prove that our 4-cylinder Chevy II is pretty competitive to the 1600, in price, economy, comfort *and* roadability. VW is the No. 2 selling car up here in the northern part of the state, you know. We'd like to set up a 2-day run, in cooperation with Volkswagen, and we'd like MOTOR TREND to act as observer."

"Okay," we agreed, "as long as we can call 'em as we see 'em."

Several days later we were in Angels Camp (if you can't find it on a map, you'll find it referred to in Mark Twain's "Jumping Frog of Calaveras County"), which was to be the start of the 409-mile run. It was to take in altitudes ranging from 4500 feet to nearly 8000 feet, until we wound up at sea level in San Francisco. Temperatures were to range from 45° down to 0°F (though that was a surprise).

The Chevy II was a stock 2-door sedan, using a 4-banger and 3-speed stick shift, while the VW 1600 fastback was a new fuel-injected job with 4-speed stick. The Chevy II was to be driven by Wilmshurst and the VW was to be handled by Herman Ultsch of Reynold C. Johnson (VW dealer in San Francisco.) Photographer Gerry Stiles and I were to follow in a backup car with a calibrated speedometer/odometer.

After topping up both gas tanks we took off together, edging out of town north on Highway 49 to Highway 50 — a road alternating from wide high-speed 4-lane asphalt to a narrow twisting 2-laner cutting through mountains capped with snow and armies of pine trees, following the course of the American River. Continuing northward, we crossed over 7382-foot Echo Summit, past Lake Tahoe to Tahoe City, 133 miles later and our afternoon lunch stop.

The cars went along perfectly from the beginning of our trek with an average speed hovering around 54 mph until we reached the 5000-foot level. From there until long before Tahoe's shores, snow covered everything and we had to put chains on the cars and drop maximum speed down to nearly 30 mph. Driving with chains didn't help our mileage and there were a number of times the cars went to lower gears just to keep from stalling while following a slow-moving pack of cars. The VW had the advantage during our crawl-session, since the car comes with a 4-speed box and intermediate gears give it an advantage over the 3-speed Chevy II. Another minus feature of the Chevy II was that after all that snow driving, there were pounds of slush stacked up on the outsides of the fenders just aft of the front and rear wheels.

Just outside of Reno we filled the gas tanks for the first time since leaving Angels Camp, bouncing the cars to remove any trapped air. We had traveled 175.5 miles and both gas gauges appeared to have stayed on "Full." On this first leg, to say we were surprised by the economy of the Chevy II (30.79 mpg), would be putting it mildly, but we were astonished by the fuel-injected VW (37.29 mpg). However, these were nothing compared to what was to come on the last leg of the trip.

That unbelievable last leg took us 185.4 miles from Reno west on Highway 50 through snow country over 7000-foot levels, down through Placerville, to the state capital, Sacramento, and on to Interstate 80 into San Francisco. Admittedly, it was not as rough a leg as the first one, lots of it being downhill and much of it no-stop roads, though we did plow through evening rush-hour traffic in

Snow packed road and watchful Highway Patrol proved VW pilot's driving expertise.

CHEVY II vs VW

Sacramento. The Chevy II's mileage went up to 33.10, but then came the shocker.

Filling the VW, it seemed the meter had just started when it stopped. Only 4.1 gallons and the tank was full, and this gave the VW a... hold on for this now.... 45.21 mpg score. What? We checked again. Okay, it's right. It's hard to believe, but we all verified it. Both machines weighed in above factory given specs, and no tampering had been done to engine or carburetion/injection sizes. Even though Dick Wilmshurst proposed and organized the test, he was the first to admit the VW 1600 had won the economy run. But, the Chevy II doesn't have to hang its head for its performance either. Besides, an interesting point we didn't think to tell him but which one should keep in mind when comparing economy figures of these two cars is this: the lower-compression VW needs the higher-priced premium fuel (unless you can find regular with an octane rating of 94 or more), while the Chevy II needs only regular fuel (generally 4c per gallon less). Therefore, on a cost per mile basis, some part of the VW's advantage would be crossed out.

Another point to keep in mind is that every comparable VW or Chevy II cannot be expected to achieve the same economy. Cars are set up differently and it also takes a conscientious driver, one who doesn't use jackrabbit starts, drives ahead to avoid panic stops, stays close to speed limits (as we did), and generally drives conservatively. The difference between driving at reasonable speeds vs. high speeds is very evident in what we achieved on the two "economy legs" and a short, "high speed" leg. This was when we wanted to check out top speed.

Though it took awhile, the Chevy II finally wrung out at just under 80 mph, while the fastback VW nicked 86 mph — sans the vibration and noise encountered with the 4-banger. After this test phase, we drove awhile longer to a gas station, where we carefully topped up again before starting on the second leg of the "economy run." We saw that we had driven 48 miles with each car. The VW dropped to 30 mpg, while the Nova dropped to a low of 17.45 mpg.

Rating the two cars in areas other than fuel economy, we would have to conclude this: the comfort and quietness of the fastback Volks are much better than the II — top speed is greater, there's more versatility with the 1600 and certainly greater fuel economy is inherent. There's more room with the Chevy II, and we like its styling better. But in its base existence, everything about it is extremely spartan. Neither car is any great shakes as a handler, so we won't rate them in this area but confine our evaluation to domestic qualities.

Our best suggestion is for Chevy II hunters to choose nothing less than a 6-cylinder engine. The little 4-cylinder is great for low-cost operation, but you can quickly forget any thoughts of performance and smoothness of operation as the quad-cylinder engine is rough at both ends of the scale — at idle and top speed. Overlooking the small engine, the Chevy II is an attractive and appealing package that comes in more combinations than any other compact car on the market today.

So, Dick, don't hang down your head. You've got plenty to be proud of, even if you did lose your own economy run. /MT

	CHEVY II	VOLKSWAGEN 1600
Total Miles Traveled	406.4 mi.	406.4 mi.
Total Amount of Fuel Used	14.05 gals.	10.4 gals.
Average Miles Per Gallon-Full Trip	28.92 mpg	39.07 mpg
Average MPG Economy Test	31.725 mpg	41.005 mpg
Best MPG	33.10 mpg	45.21 mpg
Poorest MPG (economy portion)	30.35 mpg	36.80 mpg
High-Speed (65-75 mph) Fuel Consumption	17.45 mpg	30.00 mpg
Type of Fuel Used	Regular	Premium
Fuel Tank Capacity	18 gals.	10.6 gals.
ENGINE	In-line 4-cyl, overhead valve	Horizontally-opposed 4-cyl. overhead valve
Displacement	153 cu. ins.	96.66 cu. ins.
Bore & Stroke:	3.875 x 3.25 ins.	3.37 x 2.72 ins.
Compression Ratio:	8.5:1	7.7:1
Horsepower @ RPM:	90 @ 4000	65 @ 4600
Torque @ RPM:	152 lbs.-ft. @ 2400	87 lbs.-ft. @ 2800
Carburetion:	1-bbl., downdraft	Electronically metered fuel injection
TRANSMISSION	Manual 3-speed. Column-mounted shifter. Synchro in all forward gears. Ratios: 1st, 2.85; 2nd, 1.68; 3rd, 1.00:1	Manual 4-speed. Floor mounted shifter. All forward gears synchro meshing. Ratios: 1st, 3.80; 2nd, 2.06; 3rd, 1.26; 4th, 0.89:1
FINAL DRIVE RATIO	3.08:1	4.125:1
STEERING:	Manual. Semi-reversible recirculating ball nut.	Manual. Roller-type
Gear ratio:	24:1	n.a.
Overall ratio:	28.3:1	n.a.
Wheel turns, lock-to-lock:	4.8	2.8
Turning circle:	38 ft., curb-to-curb.	36.3 ft., curb-to-curb.
BRAKES:	Drum type, 9.5 in. diameter front and rear.	Disc front, 10.9 in. and drum rear, 10.9 in.
TIRES:	7.35 x 14	6.00 x 15
SUSPENSION:	Front: independent with single lateral arm with coil spring. Rear: salisbury (one-piece unit) type axle with two single leaf springs. Double acting, direct acting shocks at each wheel.	Independent, with torsion bars. Independent, with torsion bars. Transaxle type drive. Double acting shock, telescoping type at each wheel.
Overall Length:	189.4 ins.	166.3 ins.
Overall Width	72.4 ins.	63.2 ins.
Overall Height:	54.1 ins.	58.1 ins.
Wheelbase:	111.0 ins.	94.5 ins.
Front Track:	59.0 ins.	51.6 ins.
Rear Track:	58.9 ins.	53.0 ins.
Curb Weight:	2890 lbs.	2116 lbs.
PRICES & OPTIONS	Manufacturer's suggested retail price: $2284.00	$2279.00, P.O.E. West Coast.
Automatic Transmission:	163.70	not offered
"Automatic Stick Shift":	65.00*	not offered
Overdrive Trans.:	not offered	not offered
Tires:	31.35	29.50 (whitewalls)
AM Radio:	61.10	Dealer available only
(*Torque-Drive. price is approx.)		

MOTOR TREND/JUNE 1968

Chevy II much.

Topside, it's a neat little two-door. Underneath, it's all set to move. Beefed-up suspension, wide oval red stripes and one of the greatest V8s you've ever ordered into action. It's a 350-cu.-in. 295-hp affair with 4-barrel carburetion and 2¼" dual exhausts. Nova SS. We call it Chevy II much. You'll second the motion.

Nova SS

CAR and DRIVER ROAD TEST

Chevy II Nova SS

All docile and innocent . . . the vestal virgin-image pales slightly when you turn on the engine.

Don't think that GM doesn't know about it. As a matter of fact, that's probably why the 396 is offered in the first place—it's certainly why we tested it.

Inconspicuous to the point of being invisible—that's the Chevy II. It rivals the taxi cab as the omnipresent non-car. You don't even see it in traffic unless you search it out with Chevy II radar eyes. The Chameleon II, sneaking quietly along in the curb lane in a single color, doing a terribly earnest, terribly effective job of fading into the background.

They're seldom washed, Chevy IIs, and invariably have skinny black tires with hub caps like dog feeding dishes. The car that the concept of fleet cars was made for.

If you can isolate a '68 Chevy II, say a parked one that's been washed recently, take a look at it. Surprisingly, it's not too bad. It's got the current swoopy GM lines with kind of a half-hearted fastback and what looks like too little front overhang. Something like a drag race funny car. If the one you're looking at has *lots* of overhang it's probably a Chevelle. And, in fact, that's the reason for the abrupt increase in Chevy II sales this year. Signing the check for your econo-stone is a whole lot easier if the unpracticed eye of your neighbor will categorize it one rung higher on the Chevrolet status ladder.

As enthusiasts, our interest in standard Chevy II-istry is well off the bottom of the scale. We're willing to pretend it doesn't exist if you will. But also as enthusiasts, the thought of a sleeper makes sly smiles come over our faces and our eyeballs snap both right and left in a pure reflex action to check for the fuzz. The sleeper appeals only to the most secure and sophisticated performance car fancier. There are no admiring glances from onlookers to bolster the ego. The entire driver satisfaction is based on the inward confidence that you can put the hurt on a strutting GTO or Mopar before they even realize you're a threat. Making your point in one of these street discussions by putting a fender on somebody's Super Car is pure ecstasy, particularly when you do it with an innocuous car. And to our way of thinking, a Chevy II is innocuous beyond Noah Webster's wildest dreams.

That was the plan. Order up a Chevy II with the highest output 396, the 375-hp job, and have a ball. Now, those of you familiar with Chevrolet's engine line-up are well aware that the only similarity between the 375-hp 396 and the *325*-hp 396 that was in the Camaro for the Sporty Car test (March) is the displacement. The major engine parts are all different. The 375-hp engine gets the heavy-duty block with 4-bolt main bearing caps, a forged crank with special heat treatment, connecting rods of a stronger alloy and forged, 11.0 to one compression ratio pistons. And that's only part of the story. To make the 50 additional horsepower this engine inherits Chevy's high performance cylinder heads with bigger ports and valves, a high capacity aluminum intake manifold with an 800 cfm Holley 4-bbl., and a mechanical lifter camshaft with more duration and lift.

A very serious engine to stuff into an unsuspecting Chevy II.

As you can imagine, serious Chevy IIs like this have limited appeal—in fact, no appeal at all to the normal Chevy II buy-

AUGUST 1968

er—so serious-engined Chevy IIs are not what you'd call numerous. When we asked Chevrolet for a test car they rocked back on their heels and explained that there was no way to program one into production so that we could have a new car before our deadline. If we were to have one at all, it would have to come out of their fleet. Searching turned up a cooling system test car in the engineering area which we could have as soon as its test schedule was complete. Fine with us. Detroit's Woodward Avenue is a perfect place to evaluate a sleeper so we would just stop by the Tech Center and pick up the car.

Our device turned out to be a bright red 2-door coupe with a black vinyl top. "Nova" and "SS" appeared in chrome plated script on its exterior erogenous zones. All docile and innocent we thought. Something a single working girl might own and faithfully wash every Saturday—taking *great* care about polishing the "Nova" and the "SS."

The vestal virgin-image paled slightly when we started the engine. The 375-hp Chevy IIs are built with a special low restriction dual exhaust system which has larger diameter tail pipes than the lower performance models and no resonators. That's the mechanical part, which, of course, from the outside looks no different. But the sound—there's the difference—a super low pitched rumble, sort of syncopated, that sets serious cars apart from Cadillacs and 6-cylinder stones. Every sleeper needs *that* for reassurance.

We weren't on the Motor City streets 10 minutes before we began to suspect that the Chevy II wasn't a sleeper after all. The mid-afternoon traffic included cars full of teen-agers making their way home or to work or wherever they go after school. Whatever they're doing they always keep a close watch on other cars. Every car is a potential adversary. Their whole, competitive, complicated, insecure teen-age lives require them to know every car in the world, and every engine and every combination thereof so nobody risks a run in with the heat for less than an interesting match. Big 427 Vettes will go against Hemi Mopars but ignore 390 Fairlanes completely. Older Pontiac hardtops, the Super Cars of the early Sixties, now look for other $500 cars to run with. Several times in the course of making our way across town we found ourselves first at the light beside a mag-wheeled Super Car containing two or three young males. The reaction was always the same. All eyes instantly checked out the 396 emblem beside the side marker light on the front fender. Then back to the rear wheels to see if we were running slicks. Finally a quick glance at the Chevy II's driver just to see what kind of a guy would drive such a serious car. When the light turned green the procedure was also uniform. They would gently ease off in a fashion indicating "I pass." A Hemi Dodge followed for several miles but turned off because we couldn't catch a red light. Adolescent types in the pre-drivers license age bracket walking along the sidewalk would hear the authoritative exhaust rumble, turn and focus on front fender and shout "396, 396." That kind of recognition

never happens in Fords of any kind and almost never in Plymouths or Dodges. It's the exact same response you get in a Corvette, the most tuned-in car in the U.S. Boy, were we wrong. Bolt a 396 sign on a Chevy II and it's no longer innocuous, it's no sleeper and it's not a non-car. It's a tuned-in machine. Not just with the kids, either. We were stopped by the fuzz in a residential section of Royal Oak after leaving a friend's house about 10 o'clock at night. "You look suspicious," was the explanation. "Don't drive on these streets anymore," was the warning. *They* apparently have well-defined ideas about muscular Chevy IIs also.

The final test was Ted's Drive-In, the well-known teeny-bopper, car-culture hangout on North Woodward Avenue. We rumbled down one row and after turning to go up the second the fuzz appeared and announced that if we didn't park immediately we'd have to leave.

It didn't make any difference that we were at least a generation removed from the regular customers. The car classified us. We noticed later that most other cars were allowed to drive through at least two rows before they were stopped. Now there's a clinical observation worthy of study by Chevrolet—not to mention that single working girl with her Saturday compulsion to wash cars.

Obviously, the sleeper we had expected was not only keeping *us* awake, but everybody else who saw it too. We were driving an easily recognized, big league super stocker and that's only slightly less fun. With that in mind, our scheduled test session at the drag strip was something to look forward to.

Understand that the Chevy II was a street car and not one set up for the strip. The 3.55 ratio rear gears are standard and quite a reasonable street set-up but you just get into fourth gear at the end of the quarter, which would never do for serious racers. Along with that are the hopelessly slippery UniRoyal E70-14 tires which squeal a lot but never seem to get a grip on the asphalt. You wouldn't expect a compact-sized car with an engine the size of Chevrolet's 396 to come anywhere near having acceptable weight distribution, but you're in for a surprise. Even with power steering and power brakes the Chevy II's front wheels carried only 55.1% of the total car weight—better than almost any performance car we've tested. It's all in vain, though, because with the standard tires you have to get launched so gently to avoid going up in smoke that the quarter mile elapsed times are far less than they should be. We recorded a best run of 14.5 seconds at 101.1 mph. A very impressive terminal speed—within one mph of the 427 Corvette we tested (May) but 0.4 seconds slower in ET, which is almost entirely the result of poor traction. Also interesting, from a performance point of view, is that in a series of back-to-back runs starting with a cool car the terminal speed dropped from 101 to 99 mph in four runs. To combat exhaust emission Chevrolet is now operating their engines at higher coolant temperatures and is using an air pump to promote afterburning in the exhaust manifold, both of which contribute to higher underhood temperatures and an appropriate drop in the density of inlet air as the engine warms up to operating temperature. A fresh air package

CONTINUED ON PAGE 61

... putting a fender on somebody's Super Car is pure ecstacy, particularly when you do it with an innocuous car. And to our way of thinking a Chevy II is innocuous beyond Noah Webster's wildest dreams.

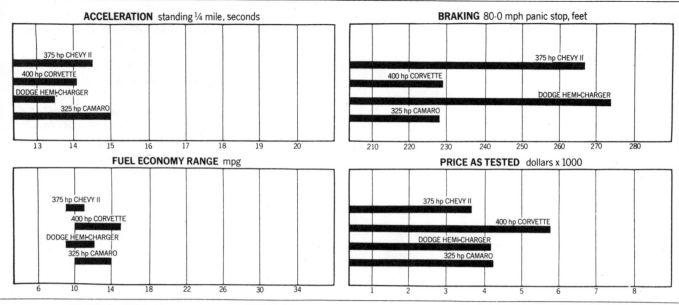

CHEVY II NOVA SS

Manufacturer: Chevrolet Motor Division
General Motors Corporation
30003 Van Dyke
Warren, Michigan 48090

Vehicle type: Front-engine, rear-wheel-drive, 5-passenger coupe

Price as tested: $3687.95
(Manufacturer's suggested retail price, including all options listed below, Federal excise tax, dealer preparation and delivery charges; does not include state and local taxes, license or freight charges)

Options on test car:
375 hp engine ($500.30), close ratio 4-speed transmission ($184.35), limited-slip differential ($42.50), power assisted disc brakes ($100.10), fast ratio power steering ($84.30), custom interior ($221.00), AM pushbutton radio ($61.10), vinyl roof ($73.75), tinted glass ($30.55).

ENGINE
Type: V-8, water-cooled cast iron block and heads, 5 main bearings
Bore x stroke .. 4.094 x 3.76 in, 103.9 x 95.5 mm
Displacement .... 396 cu in, 6500 cc
Compression ratio .... 11.0 to one
Carburetion .... 1 x 4 bbl Holley
Valve gear .... Pushrod operated overhead valves, mechanical lifters
Power (SAE) .... 375 bhp @ 5600 rpm
Torque (SAE) .... 415 lbs/ft @ 3600 rpm
Specific power output .... 0.95 bhp/cu in, 57.7 bhp/liter

DRIVE TRAIN
Transmission .... 4-speed all-synchro
Final drive ratio ...... 3.55 to one, limited slip

Gear	Ratio	Mph/1000 rpm	Max. test speed
I	2.20	8.6	55 mph (6400 rpm)
II	1.64	11.5	73 mph (6400 rpm)
III	1.27	14.9	95 mph (6400 rpm)
IV	1.00	18.9	121 mph (6400 rpm)

DIMENSIONS AND CAPACITIES
Wheelbase .... 111.0 in
Track .... F: 59.0 in, R: 58.9 in
Length .... 189.4 in
Width .... 72.4 in
Height .... 53.9 in
Ground clearance .... 5.8 in
Curb weight .... 3470 lbs
Weight distribution, F/R 55.1/44.9%
Battery capacity .... 12 volts, 61 amp/hr
Alternator cap .... 444 watts, 37 amps
Fuel capacity .... 18 gal
Oil capacity .... 4 qts
Water capacity .... 23 qts

SUSPENSION
F: Ind., unequal length wishbones, coil springs, anti-sway bar
R: Rigid axle, semi-elliptic leaf springs

STEERING
Type .... Recirculating ball, power assisted
Turns lock-to-lock 3.1
Turning circle curb to curb .... 41.5 ft

BRAKES
F: .... 11.0 in vented disc, power assist
R: ..... 9.5 x 2.00 cast iron drum, power assist

WHEELS AND TIRES
Wheel size .... 14 x 5.0-in
Wheel type .... Stamped steel, 5-bolt
Tire make and size .... UniRoyal Tiger Paw E70 x 14
Tire type .... Tubeless, 4 PR
Test inflation pressures .. F: 30 psi, R: 30 psi
Tire load rating ...... 1190 lbs per tire @ 24 psi

PERFORMANCE
Zero to   Seconds
30 mph 2.2
40 mph 3.2
50 mph 4.3
60 mph 5.9
70 mph 7.5
80 mph 9.7
90 mph 11.7
100 mph 14.3
Standing ¼-mile .... 14.5 sec @ 101.1 mph
Top speed estimated .... 121 mph
80-0 mph .... 267 ft (0.80 G)
Fuel mileage .... 9-11 mpg on premium fuel
Cruising range .... 162–198 mi

CAR and DRIVE

CHEVY II NOVA SS

CONTINUED FROM PAGE 58

like that available on the Camaro would offer a significant improvement.

The test car was equipped with Chevrolet's flawless close-ratio 4-speed transmission and their standard Muncie linkage—which is quite the opposite of flawless. Since all the enthusiasts have been setting up a unified howl to protest the inadequacy of this linkage, Chevy engineers have been tinkering with the problem and with the help of one of Roger Penske's race mechanics have come up with a compromise solution. They've found that slightly mis-adjusting the linkage makes it more difficult to get caught in the reverse crossbar on a 1-2 shift—our major complaint. On the other hand, it's harder to find reverse when you want it too, but still, no worse than a standard Volkswagen. Our test car had the benefit of this trick adjustment. It's an improvement but not the ultimate solution.

In only one area of the Chevy II's performance could we find fault, and that was braking. It was a dismal failure. Oddly enough, it isn't entirely a fault of the brakes. The leaf spring rear suspension has to share the blame. We're not categorically denouncing leaf springs—with proper development they can be made to do a more than satisfactory job. But the Chevy II needs help. During any braking test the rear axle sets up a violent hop/tramp mode which brings a certain verisimilitude to a panic stop. No matter how sophisticated the Chevy II's disc front/drum rear braking system is, controlled stops in an acceptable distance are simply out of the question. Our best stop was 267 feet at 0.80G and some stops took nearly 300 feet from 80 mph. Chevrolet is not the only manufacturer with this problem and we're sure that they're all aware that it exists. We can only shake our heads in disbelief that any responsible, government-fearing corporation would make a car with this problem available to the public.

Since our normal road course wasn't available for this test, our handling evaluation was done with the prudence required by public road driving. Once again, the leaf spring rear suspension proved to be skitterish on rough surfaces so that the need for help was obvious. However, on smooth surfaces the Chevy II is a genuinely controllable machine with none of the gross understeer found in 396 Camaros. The only suspension change for the 396 is in front spring rate which is increased to 347 pounds-per-inch from the 278 used with the 327 cube engine. You would expect this change to increase understeer but the result was in no way detrimental. Of course power induced oversteer is a readily available commodity—much like a Corvette. Chevrolet anticipated our handling evaluation by setting a noticeable amount of negative camber in the front suspension. We don't feel this is necessary for good handling in the Chevy II and even suspect that it may have been the cause of the poorer-than-normal directional stability over irregularly surfaced roads.

The optional fast ratio power steering in the test car definitely deserves mention. It provides a very quick ratio—just over three turns lock-to-lock—which would be almost unbearable in a manual steering car and at the same time has superb road feel and response. This power steering is a giant step ahead of the Ford and Chrysler competitors.

Most endearing of the Chevy II's qualities was its amazing ease of operation. Those who have envisaged the 375-hp engine to be an ill-at-ease refugee from the race track are only half right—it goes like a racer but otherwise behaves with prep school manners. It starts readily (although a little reluctantly when hot) and has no more than a pleasantly lumpy idle. The stumble we've grown to expect in big carburetor/manual transmission cars, when opening the throttle for acceleration from low engine speeds, is almost imperceptible and the properly-adjusted mechanical valve lifters are undetectable.

Driving, frequently a task in a high performance car, was nearly effortless. The difficulty with most big engine/manual transmission cars is the disparity in effort between the clutch and all the rest of the driver controls. Chevrolet is normally the worst offender in this department, not because of clutch effort higher than competition but because the other controls are far lighter. Much to our surprise, all the controls—steering, brakes, shifter, accelerator, and clutch—were commensurate in effort on our test car. The entire credit for this startling bit of coordination goes to the experimental dual plate clutch which the Chevrolet engineers had installed into the test car to get the reaction of an unbiased critic. "Power assisted clutch," is the only suitable description from a driver's point of view. It's generally understood that dual plate clutches have increased capacity and life with a smaller diameter, and at the same time less pedal effort. It's just that Chevrolet is the first manufacturer to get the bugs worked out. Obviously the manufacturing cost is higher but it makes such an improvement in effort that we think it's worth nearly as much as power steering or power brakes. Unfortunately, you can't have it just yet. It's not scheduled for production until the '69 models, and then only with 350 cu. in. 4-bbl. engines and larger.

The Chevy II's instrument panel is a masterpiece in the great Chevrolet tradition. All the instruments; the speedometer, fuel gauge and optional clock, are grouped directly in front of the driver for quick, at-a-glance viewing. Sorry, but if you want to know anything else about what's happening inside your very special 375-hp 396 (made-of-all-very-exotic-pieces), you'll have to wait until one of the warning lights comes on—which simply serves notice that you should speak of your engine in the past tense. Not only is the panel hopelessly incomplete for a high performance car, but the heater controls must be operated by feel alone. They are unlighted and completely obscured by one of the fat steering wheel spokes anyway.

Padding on the instrument panel—free from decorative ridges and amply covering the most vulnerable area—was excellent. There *are* disadvantages to ample padding and these become obvious when trying to look around the rather far windshield pillars. Trying to see out does present a problem in a Chevy II. The seating position is very high which gives a good view of the road but in effect clips off the world just above eye level. This is particularly noticeable in the front corners where the roof makes a blind spot as it dips down to blend into the windshield pillars. The same holds true in the rear quarters. Vision directly to the rear also suffers, because of the short vertical height of the sloping rear glass.

The ventilation in the Chevy II is so good it's almost a miracle. If Chevrolet was that smart, all of its cars would be this good. Apparently the shape of the body is such that the vent windows operate very effectively and yet almost totally without noise. The system is complimented by foot vents which are operated by conveniently located knobs in the side panels just below the dash. The result is ventilation way above the compact car state of the art.

Chevrolet shies away from calling the Chevy II a compact car. Rather, it's considered to be a smaller size Chevrolet that competes with Falcons and Valiants and Darts. That's about the only context in which a name like Chevy II makes any sense at all. All of this is meant, obviously, to categorize an everybody's junior Chevrolet. The junior Chevy with the senior engine, like our 375-hp Nova SS, competes with *nobody's* compact—it's an instantly recognized and feared street cleaner that pushes you well up the youthful car-culture respect poll. With the exception of the clumsy rear suspension the car is so well coordinated that we never once had the overpowered-car feeling. In fact, unless you just plain need the cargo capacity we think a performance car any bigger suffers too great a penalty in maneuverability and weight.

The 396 Chevy II sure wasn't the invisible sleeper we had expected but it was every bit as wild as we hoped. ●

CARS ROAD TEST
PHASE III SS-427 CHEVY NOVA

With stock Super-Bite suspension and optional gigonda shoes, PHASE III Nova explodes off the line with no traces of wheel hop.

Front end of test car came up a bit when blasting out of the chute. Suspension is available with 50/50 street or 90/10 strip shocks.

• **CARS**

You don't have to like its looks, but with 500 hp on tap you sure have to respect it!

IN AN ATTEMPT to break away from the ugly duckling, *el cheapo* image of the Chevy II, Chevrolet has given the car a fresh look, a new name and a total-performance option lineup. The net results of this merchandising-engineering assault is the Chevy Nova, still basically ugly but now a genuine hauler for the street-strip set.

As far as performance buffs go, the Chevy II has been just about out-to-lunch since its introduction. When it was available with the small-block hot setup engine—the 350-hp hydraulic-lifter 327—it attracted some attention from the street rats. 'Grumpy' Jenkins put it to many of the big-cube boys with his almost unbeatable 327 race machine. When this engine was dropped from the option list, the Chevy II was dropped from Hot Setup list by the SRA (Streets Rats of America).

But, like the proverbial bad penny, the Chevy II has returned. Going under the name of Nova SS, the homely little two-door coupe (hardtops not available) has staged a fantastic comeback offering the big-block Semi-Hemi 396 in 375 hp solid-lifter trim, choice of three-speed Turbo Hydro or four-speed Muncie, power disc brakes and a multi-leaf rear suspension as openers. And, if you buy one like the one shown here, you can choose from a complete page of perfor-

Status symbol side marker and light replaces the factory 396 consumer job.

Besides looking groovy, hood-mounted tach is only model that works with CD ignition.

PHASE III mill comes stock with three-barrel Holley, headers, CD ignition, special manifold and chrome competition air cleaner.

PHASE III SS-427 CHEVY NOVA

mance options ranging from a three-barrel carb to a blueprinted all-aluminum alloy ZL-1 race motor.

While the basic Chevy Nova—as is from the factory—leaves a lot to be desired, it features one redeeming factor which makes it rather appealing to a large segment of the performance market. It may have very little character, image trim or plush features, but what it does have are unreal performance and a low, low price tag. And, when you get right down to the real nitty gritty, that's what it's all about. For example, the dealer group that supplied our ultimate performance and appearance test Nova sells 396/375 models for under $2,900, SS-427 (450 hp) models for $3,695 and Phase III models loaded with every option shown on our test car for a few bucks under the magic five grand mark. You pays your money and you takes your choice!

As mentioned previously our test Nova was decked out with the full PHASE III treatment, plus whatever the dealer could find in the option list to turn it on. We would have been happy to simply test a stock 396/375 or 427/450 Nova, but they looked so ordinary that we opted for the good guy model. What almost turned us off the car was the matter of the stiff price tag which put the car into another world, defeating the purpose of the whole thing. Just for the record, however, the same performance options are available in a Plain Jane model which doesn't come with the scooped hood, scooped hood tach, functional Corvette side exhausts, vinyl roof and mag wheels.

Before even moving it outside the dealer complex we decided to go over the car with a fine-toothed comb. It's nice to know what makes a car tick, especially when it's a 3,300-pound sedan that's powered by 500-hp solid-lifter mill. Like the old 350-hp job was a bit out of hand in the traction and stopping departments, so you must think twice when you're planning on beating on a Twilight Zone machine like the PHASE III SS-427 Nova.

Once up on a lift the SS-427 Nova looks like any other SS big-block Camaro that rolls off the line. The front end sports coils and the usual jazz, while the back end sports multi-leaf springs. In charge of stopping are standard power disc brakes. Since this type of suspension is prone to wheel hop with the 396 engine, all PHASE III models are fitted with standard equipment

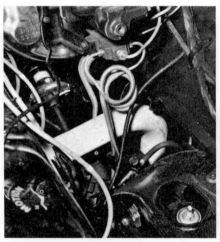
Slight cutting was necessary to route the tuned headers to the Corvette side exhausts.

All 427-cube models are delivered with beefed rears with welded spring mounts, axle tubes.

Closeup reveals welded spring perch and Super-Bite gigonda shock and anti-hop tract bar.

It doesn't take much throttle pressure to light up the Inglewood wide guys. PHASE III Nova is a genuine bear for street and strip.

• CARS

Super-Bite traction bars and four progressive sequentially-valved shocks which do a dynamite job of keeping the wheels on the tarmac. Also standard on all SS-427 models is the beefed rear which sports axle tubes welded to the center section and welded and braced spring perch mounts. This is the only setup that will live under maximum acceleration conditions.

Also visible once the car was sent up on the lift were the *standard* tuned tube headers hooked up to optional Corvette side exhausts and a *standard* Lakewood scattershield which hid the full Schiefer aluminum clutch-flywheel combo from view. With this PHASE III *stock* explosion-proof equipment you can run at any NHRA-AHRA sanctioned strip. Since the 427 engine has the same basic dimensions of the 396, the new engine fits very comfortably into the boiler room. We noticed that some minor surgery was necessary to route the tuned tubes through the side panels.

Satisfied with what was happening underneath, we gassed the brutish looking Nova up with Sunoco 260 and made tracks for the Twilight Zone. We had expected it would take some getting used to, never even thinking that since it is basically a sedan-bodied Camaro, it should react like any other PHASE III Camaro. And, that's exactly what it turned out to be. Around town it was reasonably pleasant thanks to the 4.10 Posi gears and the smooth-as-silk Hurst-Schiefer combination. Unlike the factory setup that sticks to floor when you try to powershift at redline revs, the Schiefer Rev-Lok let it happen every time we decided to power a couple of cogs at 7,000 rpm.

Because of the PHASE III standard equipment and the optional goodies, our test Nova never failed to turn heads even when we were in the company of new Vettes, Mach I Mustangs and that sort. Most supercar buffs couldn't get over the fact that someone had the guts to try and salvage a Chevy II. (The only ones that call it Chevy Nova are the factory and dealer people.) If the functionally-scooped and tach-equipped hood and SS-427 status emblems didn't get them, the rumbling Corvette side exhausts did the job. After a while, even we got used to it.

Equipped with stock manual steering, our test Nova required more effort than would be expected when parking or getting in and out of tight spaces. It is available with

Standard equipment includes headers, Schiefer-Lakewood & four-speed.

Super-Bite traction bars stiffen up suspension, dampen wheel hop.

Optional Corvette side exhausts, big tires on mag wheels and hood tach set off the stock hood scoop and SS-427 status emblems.

CARS •

PHASE III SS-427 CHEVY NOVA

Dual electric fuel pump assembly insures a constant flow to the 950 Holley three-barrel.

Clean, simple front end is set off by stock PHASE III stuff plus listed special options.

power steering, but we wouldn't advise ordering one with the quick power setup which is an additional option. We've driven prototype models and found that it was almost impossible to correct when exploding off the line, as normal fishtailing and torque-producing sway can be turned into a dangerous lane-changing affair if one corrects just a bit too much. It's just too sensitive for a car loaded with this much solid-lifter torque.

Because of the unreal gobs and gobs of torque available with the 427 engine fitted with standard equipment 950-cfm Holley three-barrel, special aluminum high-rise manifold with plenum chamber, complete Super-Spark capacitive discharge ignition with Ramcharger wires and a dual electric fuel pump (mounted in the trunk), we found the super wide 8-1/2-inch Inglewood tires almost inadequate for hard-core burn outs. The car comes out straight and the bars and shocks completely eliminate wheel hop, but the tires do a lot of

• **CARS**

1969 CHEVROLET SS-427 NOVA SPECIFICATIONS

ENGINE
Type	OHV V-8
Displacement	427 cubic-inches
Compression Ratio	11-to-1
Carburetion	950 CFM Holley Three Barrel
Camshaft	Solid, 560-inch lift
Horsepower	500-plus hp
Torque	N/A
Exhaust	Tube Headers, dual pipes
Ignition	Mallory-Motion

TRANSMISSION
Make	Four-speed, Muncie
Control	Hurst Floor shift

REAR END
Type	HD Positraction
Ratio	4.10-to-1

BRAKES
Front	11.0-inch power-assisted discs
Rear	9.5-inch power-assisted drums

SUSPENSION
Front	Beefed HD independent coil springs
Rear	HD Multi-leaf, traction bars
Steering	Manual
Overall Ratio	28-to-1

GENERAL
List Price	$3,695.88
Price As Tested	$5,200.00
Weight	3,3180
Wheelbase	108 inches
Overall Length	184 inches
Tire Size	8½" Inglewood

PERFORMANCE
0 to 30 mph	2.8 seconds
0 to 60 mph	5.1 seconds
Standing ¼ mile	115 mph
Elapsed Time	11.88 seconds
Top Speed	125 mph (EST)
Fuel Consumption	6-11 mpg

burning. This condition is almost non-existent on a stock 427 model fitted with Turbo Hydro (which is available on all stock and PHASE III 396 and 427 Novas). The power disc binders offer the needed whoa to match the go.

The SS-427 Nova will never win any handling awards but it offers all the typical purchaser of one of these brutes, is looking for. The Super-Bite stuff stiffens up the suspension giving you a good feel of the road, but the car definitely plows some when putting it through the paces. The same exact Camaro package handles better and in general is easier to control under full power conditions. As far as acceleration, passing power and overall throttle response go, there's no other production car that can touch it. With slicks, 4.56 gears and the optional PHASE III cam, it's a mid-11-second, 120 mph machine. As tested with slicks and 4.10 gears it was good for high 11's.

Even though the SS-427 Nova is not as attractive as a comparable Camaro, it does offer a basic cash savings plus the fact that you can put real-sized people in the back seat and real-size luggage in the trunk. And, there are many people who can use the extra bucks and the space as well.

After a while the PHASE III Nova started to grow on us. The tough stuff goodies more than made up for the lack of styling class and the sound and the fury of the big-guy motor put the Nova into the big league. Four days after we returned it to the Bladwin-Motion Performance Group, it was sold to a man who traded in his Camaro because it was too small for the family.

'69 Chevy Nova SS. Up and at 'em!

Louvers on the front fenders, a bulging hood and a throbbing exhaust note let people know this one's no imitation.

Backing up a standard 300-hp V8 is a muscular foundation: special suspension, an extra-tough clutch, red stripe wide ovals on 7" wide wheels, a special 3-speed and power disc brakes.

The '69 Chevy Nova SS, the car that woke up swinging.

CHEVROLET

Putting you first, keeps us first.

CHEVY'S SECRET WEAPON

Unless you look at it closely you are in for a surprise.

For many years hot-rodders have been sticking big, hairy V-8 engines in light little cars. Sometimes it works and sometimes the whole package comes unglued. This year the Establishment at GM is trying it with the formerly mild Chevy II Nova, now called simply the Chevy Nova, with quite definite success.

Last year the Novas came with three engine options, the most powerful of which was a 307-cubic-inch, 200-horsepower V-8. This is a lot of horsepower to the import buyer, but pretty limp-wrist for the Detroit Iron fans. For 1969 the three old engines are still there, but Chevy has added three more: a 255-hp/350 cu. in. regular gas V-8, a 300-hp/350 V-8, and the big 350-hp/396 cu. in. honker that we had in our test car. The last two engines are only options for the Nova SS Coupe version, with both the 'SS' and the '396' swapped from last year's successful Chevelle 396 SS model.

With nearly double the horsepower (from 200 to 350) from last year, the Nova becomes quite a bomb, using the term in the good automotive, not the bad theatrical sense. And even with this beast of an engine, the six-inch wide tread gets the power to the ground without a chirp, other than on foot-to-the-firewall starts. Cruising the freeways at 50 mph, a jab at the throttle will take the SS to 65 in three seconds flat, with everything feeling stable and smugly satisfying.

The newest gimmick on the Chevy line for '69 is the steering column anti-theft lock. The process of setting and defusing varies with different cars, so each takes a little time to get used to. On the Sting Ray the transmission must be in Reverse to get the key out, and in Neutral with the clutch in to start. With the 1969 Nova the automatic must be in Park for locking, and either Park or Neutral for starting. Once you do have it in Park, and turn the key to lock, the steering wheel will move only a few degrees before it clunks into a locked position, and the transmission won't come out of Park.

Now, a good locksmith could defeat the whole system, or a crook with a tow truck and rear-wheel dolly could cart the Nova away. This is not the purpose of the anti-theft lock. Very few car thieves are expert locksmiths or own a tow truck. Most steal cars they find with the key still in the ignition. If you're going to commit this common oversight, don't be too surprised when your car is heisted. The Nova does have a buzzer to warn you anytime you open the door and the key is still in the ignition lock. Some owners will undoubtedly

ROAD TEST

find it annoying and disconnect it. The system is not fool-proof.

The basic 111-inch wheelbase Nova is the same as it was in 1968, with a slight change in exterior trim (the factory calls it 'refined'), but it is a very good basic design. The Nova is low and sleek, and the Wide Ovals make it look powerful and racy. As in most fastbacks, all this sleekness is gained at the cost of headroom and legroom in the rear, but it's not bad if you think of the car as a two-plus-two. Hopefully two of the four are children or midgets.

It would be a pleasant novelty if American automobile manufacturers could leave a clean design alone, but they rarely can resist the urge to tinker, to 'refine.' As the Nova 396 SS is a powerful machine, they must clutter the exterior with 'styling items' they feel represent power and speed. The Nova SS coupes come with two 'simulated air intakes' on the hood, and 'simulated front fender louvers' on the sides. We can only assume that their purpose is to take in simulated air. In actuality they do have a function: when you wash the car, they retain a half-cup of water which runs down the hood after you have finished drying it.

Power train

The Nova SS Coupe is available with two engines, as we mentioned, the 300-horsepower 350-cu.-in. Turbo-Fire V-8, and the big 350-horsepower 396-cu.-in. V-8 that we had in our test car. The 396 is the same as in the 1969 Chevelle, with a four-barrel Rochester downdraft carburetor, with 1 3/8 inch primary and 2 1/4 inch secondary bores, dumping the fuel-air mixture into the 10.25:1 compression ratio cylinders. This means high-octane gasoline, of course.

The advertised horsepower is 350 at 5200 rpm, and torque is 415 lb./ft. at 3400 rpm. While these factory figures may not be exact, the 396 engine certainly does supply a big batch of horses and pound-feet for a coupe weighing in at 3200 pounds. Remember that a Pontiac Catalina will tip the scales at close to 4500 pounds.

The transmission in our Nova SS was a heavy-duty three-speed "Turbo Hydra-Matic" with a 2.42:1 low gear ratio. The normal axle ratio is 3.55, with three optional ratios: 3.31, 3.71, and 4.10.

The 396 engine has a cast iron block and heads, with domed cast aluminum pistons. The piston pins are chrome steel, the connecting rods drop-forged steel, and the crankshaft is forged steel rather than the cast nodular iron of the smaller and low-

THE NOVA SS

Ammeter, water temperature, fuel, and oil pressure gauges are on a console above shift lever of 3-speed automatic.

Dashboard has tach, speedometer, and clock in front of driver with warning lights for high-beam, parking brake, and low fuel. Lower switches are (left to right) lights, wipers, AM/FM radio. Heater-demister controls at right of wheel.

Chevy Nova 396 SS has Wide Oval tires and smooth, single-headlight lines. Only the overly large chrome bumper destroys the otherwise good lines.

Front disc and rear drum brakes pull the Nova 396 SS down quickly, pulling 26-28 . per sec.² on wet asphalt, wit minor dive.

Cornering is smooth with Nova, wide tires and Camaro-like suspension work together to make understeer easily controllable. Power is on at this time.

er-performance Chevy V-8s. The chain-driven, cast iron camshaft runs in five steel-backed babbit bearings, with quite high lobe lifts: .4614-inch for the intake valves and .4800-inch for the exhaust valves, both with the hydraulic lifters at zero lash. Spring dampers are used inside the valve springs in all cases.

Performance and handling

The big, fat tires on the SS are Firestone polyester cord Wide Oval E70-14s. These are standard on the SS and optional on the non-SS Novas. Mounted on 7x14 wheels, these weinies put a six-inch wide tread pattern on the ground. The trouble is that is still isn't possible to get all the 396 engine's power to the ground on really pedal-stomping starts.

At the Orange County Drag Strip we tried it, and lost several seconds each time as the tires spun loose and the Nova slithered forward. Feathering the throttle finally got the tires to grab, about two car lengths from the timing light, and then the acceleration was truly impressive. Our best time for the quarter-mile, on a damp asphalt strip, was 17.1 seconds, with a trap time of 91 mph at the end.

On the freeway the tires have a fine grip on the pavement, and satisfying bursts of acceleration seem almost required. Rolling along at 30 mph, a

floor-boarded throttle gets you to 65 mph in 6.2 seconds, from 40 to 65 in 4.7 seconds, and from 50 to 65 in 3.0 seconds.

Suspension on the Nova SS is nothing exotic, but it works. This is the most important thing. The front uses independent coil springs and shocks, with multi-leaf springs in the rear. Non-SS Novas use a single rear leaf. The rear shock absorbers are bias-mounted, with the right shock ahead of the axle, the left behind.

We drove the SS for two weeks, making regular freeway runs with all their hazards, plus a couple of fast trips over mountain roads. We batted through curves at high speed, and we pulled some pretty quick avoiding actions on the freeways. The SS always felt low and secure. Part of this feeling is due to the wide tires, part to the Camaro-like suspension, and part to the 350 horses that lie under your right foot.

The first time that we filled the 20 gallon tank, we mentally figured our gas mileage would be in the 10-11 miles per gallon range. We don't try to be a feather-foot when we test a car, and the gas mileage quite naturally suffers. We were pleasantly surprised when we added up our credit tabs after a few tankfulls. We averaged 13.3 miles per gallon, which is not far off the supposed national average of 14.

With fewer Stop Light Grand Prix show-off starts the SS might well get 15-16 miles per gallon. There's only one trouble. With the deep rumble from the exhaust, and all that torque,

ROAD TEST

Relationship of throttle to brake pedal is good. Discs at front provide effective braking.

it's not easy to be a balloon-foot. It's far too much fun to pretend that every traffic light is a drag race Christmas Tree. You try to hit green light right on the nose, as the track officials in their black-and-whites are quite hard-nosed about red-lighting a start. You come down fast and gentle on the pedal, keeping the skins just short of squealing, and in about four seconds you back off at a neat 35 mph, all quite legal and square. Then you glance in your rearview mirror at all the sleds still plowing across the intersection, and you smile.

Brakes and safety

The non-SS Novas come with drum brakes all around, with front discs an option. Although the front drums are finned for better heat radiation, the discs are the only way to go. Or, more correctly, to stop. The SS coupes come with front discs as a regular feature. With the SS you know you're going to drive it hard, and fling it around in a sporting manner, so the discs' lack of heat-fade and their not being affected by water will add both to safe driving and to peace of mind.

The rear drums on all Novas are 9.5-inch diameter, with molded asbestos linings. The hydraulic system has dual master cylinders, with a warning light if the pressure in either system drops and with corrosion-resistant brake lines.

The brakes on our test car were very good under all conditions, dry or wet. At the drag strip we made a series of semi-panic stops from 60 miles an hour. We tried to get maximum brake effect without breaking the tires loose from the asphalt. This is not easy to do with power brakes and with a wet drag strip. The best we could decelerate was at between 24 and 26 feet per sec^2 or from .76 to .82 gravity deceleration, which is pretty damned good. We feel that the figures would have been at least another couple of ft/sec/sec better on dry pavement. The wide footprints of the E70-14 Firestones let the brakes themselves do more work before the

MAY 1969

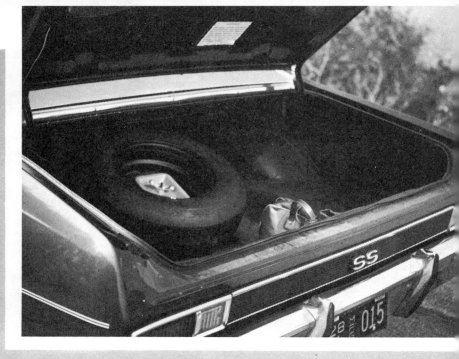

Trunk is not huge, but is more than adequate for most uses. Fat spare tire in dead center negates much of the trunk's usefulness for large suitcases.

point is reached at which the tread breaks its adhesion with the ground.

With the SS hitting just over 90 mph at the end of the quarter-mile strip, we never did get a chance to see if the 120-mph speedometer was for real. We seriously doubt it, but 100 mph could be easily expected, and nudging 110 on the way to Las Vegas is conceivable. That big four-barrel Rochester can shove a lot of air and fuel into 396 cubic inches!

Comfort and convenience

The individual Strato-Bucket seats in the test car were very comfortable, even for long trips, and when adjusted for our six-foot frame, gave a sporty and effective arms-out driving position. The two-inch seat belts seem poorly planned, and are one thing we would quickly replace. There is no ratchet on the door-side tab end, so the buckle ends up on the driver's left hip bone. We feel that the ultimate in both safety and comfort would be three-inch competition belts, with their big, quick release buckle right at dead center.

The standard GMC shoulder belt is still a jury-rigged, two-inch strap bolted to the roof far behind the driver and passenger's heads. The tab end of this snaps into a second seat belt strap and buckle that keeps getting mixed in with the regular lap belt straps and buckles. The shoulder belt would keep you from being hurled into the steering wheel, but if you were violently flung to the left, and particularly if the door came open, you'd probably get it right in the neck.

General Motors still clings to ignition and trunk keys with the notching on only one side. Now, admittedly, this is nit-picking, but when you have to fumble around on a dark rainy night, trying to get the dumb door unlocked, it can be annoying. The majority of non-GMC manufacturers long ago went to double-faced keys, which work whichever way you jam them into the slot. We are aware that all that is required with a single-faced key is that you take a look at it, or feel for the notched side. This is fine, except at night, wearing gloves.

We found the Nova SS coupe to be completely dry when driving through heavy rain storms, and the two-speed wipers worked fine for heavy or light showers. Only once, after the Nova had been parked all night in a downpour, we found a small puddle on the passenger's floor mat. We never did figure out where it came from. Perhaps it was only simulated water that leaks in through the simulated air intakes on the hood.

It is nice to have a fairly full set of gauges on the Nova, rather than the row of idiot lights for which Detroit is so famous. Unfortunately the only ones that are on the padded dash, in front of the driver where they can readily be seen, are the tachometer, the speedometer-odometer, and the clock. The engine temperature, oil pressure, fuel level, and ammeter gauges are on an extra-cost console ahead of the shift lever, and completely out of sight under normal driving conditions. It's not a lot of trouble to lean over to the right to see them, but it would be a lot safer if they were visible at all times, so the driver's eyes could flick across them from time to time as a fast condition check.

Driver vision is good in the Nova coupe, even with the headrests on the two front seats. Before pulling a lane change to the right, a quick glance over the shoulder gives you a clear view of the truck in your rearview mirror's blind spot. The side mirror, adjustable from inside, is very close to the driver and doesn't give a very wide field of view. Our preference would be for a set of racing mirrors out on the front fenders. Without changing your field of vision you would have a constant check on whatever was coming up from behind, and on both sides.

Dollars and cents

The base price on a Nova Coupe is $2389, but you might as well forget this figure, as it has little meaning towards the Nova 396 SS we tested, or to any other Nova you'll see on the street. The twenty-three 'extras' on our test car added $1763.15 to the base price. These included Soft-Ray tinted glass ($32.65), Strato-Bucket front seats (an unbelievable $231.75), head restraints ($15.80 more), remote control mirror, center console, Positraction rear axle, the 396 optional engine ($184.35), power steer-

The 396-cu.-in. V8 in the new Nova is the 350-horsepower Chevelle engine, gives the Nova nearly double top horsepower for last year, makes it a real swinging two-door for any use.

ing, wheel covers, heavy-duty battery, AM/FM stereo radio ($239.10 right there), Turbo Hydra-Matic transmission ($221.80), bumper guards, Custom Exterior (whatever it means, it adds $97.95 to the tab), and "Nova SS Equipment" at another $280.00.

This takes our $2400 fastback up to $4169.65 delivered on the West Coast. Add sales tax and license, of course, taking it to around $4500 by the time you drive it off the dealer's lot. That's for cash; if you finance it, add the bank's charges.

Summary

We were ready in advance to dislike the new Nova 396 SS. Why should Chevy stuff such a big engine into a relatively light body? Is it some sort of gag? We very quickly found out that it isn't a gag. It doesn't come all unglued. It is a very workable and sporty combination of engine and body. By the end of our test we were really very unhappy to give the two-door back. With many cars we can't give them back fast enough.

Oh, not that we wouldn't make a few changes if we owned one. We'd put in competition seat belts. We'd pull off the fake air intakes. We'd throw away the fake mag wheel covers and get some real maggies. But that's about it.

What more can we say about the Nova 396 SS coupe? We liked it very much. Its body lines are simple, smooth, and uncluttered without being a crowd-stopper. The 396 engine gives it more than adequate power for any type of driving, with fuel mileage that is reasonable. After all, you can't expect a Beetle's gas mileage with 350 horses in your stable.

The Nova SS is the type of enjoyable car that is annoying. If you owned one, you'd think about it out in the parking lot while you sat at your desk. You'd think about firing it up . . . baroom-baroom . . . and heading for the nearest long and winding road. ♠

Chevy Nova 396

Data in Brief

DIMENSIONS

Overall length (in.) 189.4
Wheelbase (in.) 111.0
Width (in.) 72.4
Height (in.) 52.5
Track (front) (in.) 59.0
Track (rear) (in.) 58.9
Weight (as tested lb.) 3373
Fuel tank capacity (gal.) 20.0
Luggage capacity (cu. ft.) 13.8
Turning diameter (ft. wall to wall) 40.9

ENGINE

Type . . . . V-8 ohv
Displacement (cu. in.) 396
Horsepower (at 5200 rpm) 350
Torque (at 3400 rpm) 415
Fuel required . . . . premium

TIRES AND BRAKES

Tires . . . . E70-14 Firestone
Brakes (front) . . . . disc 11 in.
Brakes (rear) . . . . drum 9.5 in.

SUSPENSION

Front . . . . Unequal A-arms with
 coil springs/shocks
Rear . . . . Salisbury rear axle with
 multiple leaf springs

PERFORMANCE

Standing ¼ mile (sec.) 17.1
Speed at end of ¼ mile 91.0
Braking from 60 mph (ft.) 148.9

PASSING

30 to 65 mph (sec.) 6.2
40 to 65 mph (sec.) 4.7
50 to 65 mph (sec.) 3.0

13-SECOND GROCERY GETTER

By Steve Kelly ■ A Chevy II is a neat car to drive even if it doesn't have a 396 V8 with 375 horsepower running through a Turbo Hydra-Matic. Last year in Detroit, someone turned us on to a "special" Z/28-outfitted Chevy II which was almost as good as round tires. It was great. So this year we do a session in an automatic-transmissioned coupe with a monster motor, and it positively qualifies as the boss of low-cost supercars. This neon-green two-door tagged out at $3800 by the time it reached us, but it could've been down to $3100 by forsaking power steering, radio, and a couple of other add-ons.

An odd thing is revealed when trying to get sales information for a 396/375-horsepower Nova: It isn't listed in any of the Chevy sales literature. The 375-horse engine is conveniently omitted from every listing in salesroom brochures. But the thing is there. A 375 engine is an L-78 option which costs over $300 extra on a Chevy II. The Turbo Hydro is an M40. The Nova is a model 11427. Chevrolet may not admit they build the engine to general audiences, but every informed hot rodder knows about it. NHRA knows about it too, and the combination Chevy II and 396 engine is stock in their book.

Chevy II's are very Camaro-ish all over their 3500-pound structure. A platform floor and forward subframe are identical on both cars. Front and rear suspensions are the same, including absence of U-bolts for spring-to-axle attachment. A single bolt passes through the multi-leaf spring to the perch flange on the housing. That's got to go right now if any high-speed or high-torque running is to be done. They also have staggered shocks; the right one is ahead of the axle, while the left is aft of the axle. Camaros have 108-inch wheelbases, and Chevy II's have 111-inch measurements, with the added inches found in the rear seat compartment, allowing rear passengers at least an even chance at moderate comfort.

This car is a compact by description and generally fits the size guidelines. It can be had with a four-cylinder engine, though it's hard to see why. Other engines consist of sixes and small-block eights. The trunk compartment is adequate, there's great engine compartment space, and sheet metal is pleasingly formed. It could use more sound deadening, and the front end is noticeably overweight with the 396 block aboard.

It needs power front disc brakes and gets them; because the Nova SS optional equipment, which includes discs, is a mandatory item with big-engine Chevy II's. The SS option costs $280.20.

The Turbo Hydro does its job well and is a more sensible setup for the street than a four-speed, especially with this powerplant. It didn't grow overly warm in traffic but it did gobble fuel. Average reading was under 10 mpg, and this was aggravated by not being able to get the tank full. It's supposed to hold 18 gallons but 14 was usually the maximum because the spout is placed low (behind the license plate), and gas backs out the tube before the top of the tank is wet.

Quarter-mile tryouts at Orange County Raceway were gratifying, even though the car didn't show full strength. It consistently bested the 103-mph mark, reaching 105-plus for a top record. A low e.t. of 13.87 seconds couldn't be duplicated or bettered, and the very worst e.t. was 14.47. Average time over the 1320 was 14.228 seconds. The best time was done with the column-mounted shifter in "D." That proved to be the smarter move, as the transmission is scheduled to shift by itself right at the top of the power curve, which is 640

HOT ROD • JULY 1969

74

Big-block Chevy II's take some searching to find in a dealer's book, and for that matter, they're rather hard to catch in a quarter-mile

photography: Gerry Stiles and Steve Kelly

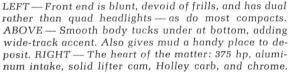

LEFT — Front end is blunt, devoid of frills, and has dual rather than quad headlights — as do most compacts. ABOVE — Smooth body tucks under at bottom, adding wide-track accent. Also gives mud a handy place to deposit. RIGHT — The heart of the matter: 375 hp, aluminum intake, solid lifter cam, Holley carb, and chrome.

m, and the trans knows this better an the driver. A little anticipation is eeded on the shift, somewhere between 200 and 300 rpm. You've got to good to get that accurate, and while seems too easy to let the shifts happen automatically, that's the method for ort time on the clocks.

The stock E70-14 Uniroyals stayed on d proved best for the conditions after we tried a pair of Casler-capped oodyears. The seven-inch-wide strip es were too tall for the rather high ar. The initial request was for a 4.10:1 g, but a 3.55:1 with positraction was nt. The UniRoyal wide-pattern tire isn't actly the ultimate drag tire going, but nce we had an automatic with low engine stall speed, and because the other bber cut top end speed by over 500 m, they got used heavily with 32 psi. hile the 4.10 gear would've blown mile e figures out of reason, the resultant od acceleration times would be worth riting about.

It takes only two runs to heat the 396 to the point where performance is ndered. A complete cool-down is needed, and thermostat removal is a wise ove. A seven-blade fan which doesn't ve a viscous clutch is fitted. The fan makes a tremendous amount of noise at high speed, so while it does move a lot of air, this same action gives out the bothersome noise level. A Flex-A-Lite fiberglass fan was sent to us too late to fit on the Chevy II, and while it would've been interesting to check the cooling power of it versus the stock item, we'll at least have the fiberglass one for use on a later test car.

Wheel hop wasn't evident at all, but it's likely that a higher stall speed on the Hydro converter, and some good traction tires and low gear, would produce traces of the unwanted bounce. Weight transfer is very noticeable on initial acceleration, due mostly to the height of the relatively short-length car. The transfer may not be as severe as it feels, but the point is it feels pretty healthy. Stall speed with the stock transmission is 1000 rpm. Anything more than that gets the rear wheels creeping over the rear drum brakes, and the front discs alone won't hold the car at rest.

There's a pretty empty feeling just as the car leaves full stop, and it takes a second for the engine to recover and be able to pull. As the gearbox upshifts to second, the stock tires really let go; and 20 feet of rubber is not an uncommon souvenir to leave on the track. The 2-3 upshift is just as speedy as the 1-2, but car weight and engine speed don't allow the rear wheels to break loose. That 1-2 rear wheel traction loss could be cause for a .05-.10-second time loss. A firm and straight hold on the steering wheel is mandatory, especially if the driver isn't used to the car. It's doubtful the car would get far out of shape, but a surprise like that to a fresh pilot might cause any number of different — and odd — reactions.

For the driver who knows what he's doing and likes to handle a potent car, this little setup is definitely enjoyable. One refreshing aspect is there's an almost total lack of identification on the outside: no phony scoops, stripes, bumps, badges, numbers, or names. There are a pair of quasi-scoops on the hood that resemble a running board grating more than a scoop, and the engine size is noted on the front sidelight trimpiece, but that's all. The other guy doesn't figure out there's a 375-horse V8 under the running board grates until he puts his ears near the tailpipe ends. That's a full story right there; what a delightful resonance. The sleeper car era has returned.

JULY 1969 • HOT ROD

13-SECOND GROCERY GETTER

LEFT — Oval-shaped wheel has horn buttons at each end of center spoke. Tachometer option can be researched, or an RAC "nine-grand" unit like this one fits neatly on column in good view. Radio is directly to right of steering column. RIGHT — Trunk measures out at 13.8 cu. ft., third largest in Chevy line-up. Deck lid opening allows full access. Front and rear overhang is smaller than on 3-inch-shorter Camaro.

A tachometer may be fitted in the well-thought-out and -designed dashboard, but only when a full array of gauges is ordered with a center console, and with bucket seats. The instruments other than tachometer nest atop the $53.75 console and consist of oil pressure, ammeter, fuel gauge (the tach displaces it from the dash face), and clock. These are well out of normal viewing range and cost $94.80 complete. The tach, though, is in a good spot for quick reading. The bucket seat addition runs $231 and some change and is part of a Custom Interior package which really does help appearances. So if a tachometer really is what a buyer is seeking, the total tab gets over $350, which is overstating the case a bit but does point out assembly line and bookkeeping hang-ups involved with new car ordering. Since our test car was devoid of such cost-building options, we fitted a transistorized RAC tachometer to the steering column, adjusted it for quick reading, and considered that the savings was something like $300. Since accessory tachometers are usually much more legible and more closely calibrated, this is the way to go for satisfaction in driving and to save money.

The bench seat in this car seemed more comfortable than bucket seat models, at least to us, and was quite a bit more practical. The seat cushion could be longer, or maybe this wouldn't be needed if seat travel was extended more to the rear. Camaros and Firebirds have this problem too, and we suspect the seat adjustment is the same unit on all three. Someone at Fisher Body must figure back seat room will look more appealing if the front seats aren't allowed to go very far back. That may be, but few people put a measuring tape to the test. Common sense provides that a driver isn't going to crush his rear seat passengers' legs against the front seat back, but he will extend the seat fully aft if there's nobody in the way. Mustangs and Cougars can have their front seats almost touching the rear seat cushion, if needed. That better idea isn't licensed.

Seating is high in both front and back. An over-six-footer is likely to plant his hair against the headliner in the back seat, which is okay because it's washable; but front seat vertical height is ample to the six-foot-four-inch level. Almost everyone will bump his head on the roof sill when getting aboard, as the door top opening isn't high enough to match the seating height, if that makes sense. What happens is that your head is last in, and it can meet the metal on the way.

Rolling the windows up tight seals out road noise, but the fan song is still able to penetrate. Smaller-sized engines and/or ones with clutch release fans should provide quiet running. The center-top-mounted rear view mirror suffers from vibrations at speeds over 55 mph. The vibration is enough to effectively blur sharp images of cars following.

It's plain to see we dug up flaws in the Chevy II design, most of them not incurable. None of them are heavy enough in nature to discount the car from contention as a bargain for super-cars. Nova SS coupes with 396 engines are populating every area of the country where performance-conscious car nuts exist. Chevrolet certainly doesn't flaunt the Chevy II 396 package, and the 375-horsepower listing is harder to find than a tax deduction; but the search for it all is certainly worth the end result. Keep the rear bumper shiny; it's liable to be seen by a lot of unsuccessful competitors.

VEHICLE
Chevy II Nova SS coupe

PRICE
Base $2405.00
As tested $3829.10

ENGINE
Type V8
Cylinders 8
Bore & stroke .... 4.0938 x 3.76 in.
Displacement .... 396 cu. in.
Compression ratio 11.0:1
Horsepower .... 375 @ 5600 rpm
Torque .... 415 lbs.-ft. @ 3600 rpm
Valves: Intake .... 2.065-in. dia.
      Exhaust .... 1.720-in. dia.
Camshaft:
  Lift .... .518-in., net lift.
  Duration ...... Intake, 282°; exhaust, 285° (actual duration, measured from starting point .020 up ramp)
  Overlap ..... 55°, actual — measured same as above
Tappets .... Mechanical, .030-in. lash
Rocker arm ratio 1.70:1
Carburetion ...... Single Holley 4150 4-bbl
Exhaust system .... Dual with single dual-input muffler; 2.25-in.-dia. main and tailpipe

TRANSMISSION
Type .... Three-speed automatic; Turbo Hydra-Matic 400. 3-element hydraulic torque converter and a compound planetary gear set. Column lever
Ratios: 1st 2.52:1
      2nd 1.52:1
      3rd 1.00:1

DIFFERENTIAL
Type .... Semi-floating axles, overhung pinion gear; single-unit "live" axle housing. Limited slip
Ring gear diameter .... 8.875 in.
Final drive ratio 3.55:1

BRAKES
Type .... Front disc/rear drum with integral Delco Moraine vacuum booster
Dimensions: Front disc .... 11.0-in. dia.
      Rear drum .... 9.5-in. dia.
Total effective area .... 114.0 sq. in.
Percent brake effectiveness, front .. 64%

SUSPENSION
Front ..... Independent Short-Long-Arm type with coil springs mounted atop each lower control arm
Rear ...... Multi-leaf spring-suspended single-unit axle housing with staggered shock (right ahead of axle, left aft of housing) setup for axle windup absorption
Shocks ... Direct, double-acting tubular. 1.00-inch piston diameter
Stabilizer .... Front only, link type, .687-in. dia.
Tires .... E70-14 Uniroyal
Wheel rim width .... 7 in.
Steering:
  Type .... Saginaw semi-reversible recirculating ball stud with integral vane-type power-assist pump
  Gear ratio 17.5:1
  Overall ratio 20.7:1
  Wheel diameter .... Oval, 16.25 x 15.50 in.
  Turns lock to lock 3.5
  Turning dia., curb to curb .... Not available

PERFORMANCE
Standing start quarter-mile (best)
.... 13.87 sec., 105.14 mph

FUEL CONSUMPTION
Best reading .... 11.02 mpg
Poorest .... 8.44 mpg
Average .... 9.94 mpg
Recommended fuel .... Premium

DIMENSIONS
Wheelbase .... 111.0 in.
Front track .... 59.0 in.
Rear track .... 58.9 in.
Overall height .... 52.4 in.
Overall width .... 72.4 in.
Overall length .... 187.4 in.
Test weight .... 3570 lb.
Body/frame construction .. Combination, integral body-frame. Separate forward ladder sub-frame
Fuel tank capacity .... 18 gal.

HOT ROD • JULY 1969

CAR LIFE ROAD TEST

NOVA SS350

The last of the optioned Compacts has sporting performance in a family car. It may not be with us much longer.

Do YOU SUPPOSE Chevrolet's Nova appreciates the promotion? The Nova began as a stepchild. Y'mean they don't want an American Volkswagen? They're buying shrunken Fords? So we'll give 'em a shrunken Chevrolet, and we'll call it Chevy II.

Then an orphan. Last of the fancy compacts. Corvair, Falcon, Comet, Rambler, all dead. Tempest, F-85 and Special, grown up and gone away. Valiant? Packed off to the workhouse, or dressed up with a new roof.

And now, without ever being a member of the establishment, the Nova is defender of the status quo. Promoted in the field, so to speak.

The Imports have attacked in force, and the wagons are drawn up in a circle. (The Aborigines are on the inside, for once.) The other tribes, Maverick and Hornet, are out there circling with the attackers. So are the mercenaries, Opel, Cortina and so forth. And we know which side Chrysler's 25-Car will take, don't we?

Here's a trim package, Nova, and a special suspension, and a rapid-firing engine. Hold this line if it takes all summer, and when the XP-887 gets here, we'll put the devils to rout. And

FRONT DISC BRAKES are a mandatory feature of the Nova SS package, and they are a good thing to have. The test sequence showed only slight fade. Front/rear proportioning could have been better, and stops weren't as short as we'd like.

NOVA

you'll get a medal. (Posthumously, of course.)

This gallant warrior is a square Ponycar, and you can take that any way you choose. The Nova is the base from which Camaro sprang, with a unit body and front sub-frames, conventional front suspension, leaf springs and live axle in back. The test car had the highest-tune 350 listed for the car, the 300-bhp model also sold in Camaros, Corvettes and so on. There may still be some of those mysterious order blanks with 396s and such listed, but we're told the 350 is as much as the factory will offer this year.

The SS package is just that. With the 350 comes either a four-speed manual or a Turbo Hydra-Matic automatic transmission. Power-assisted disc brakes are included, which may be Chevrolet's way of working itself into a policy of discs for everybody. And stiffer springs and firmer shock absorbers, wheels with 7-in. rims (very wide for a car of this size) with belted/bias E70 tires, and the usual collection of trims, letters, badges and brightwork.

The 300-bhp engine is not a performance engine. It moved the car along with the traffic, maybe a bit in front of the herd, but it's no threat at the strip. The gearing was chosen for highway driving, and the test Nova did not leap off the line. But it was quiet, and it should give good service for a long time.

What illusion of performance there was came from the transmission. The factory did right in not listing the two-speed Powerglide for the Nova SS. (When they push Nova off the plank, we hope they've tied the Powerglide to it.) The three-speed shifts down so quickly and easily that the driver has all available power on tap when he needs it. The gears work out well; the power comes on in second gear for passing and merging, which is where the driver of such a car will need it.

He will also welcome some economy. Yes, good miles per gallon, a subject not usually dwelt on in these pages. Our test circuit result was 13.2 mpg, about average for the test car's engine and weight. But one staffer took the car on an extended trip up the California coast, thinking of all those curves and scenery through the mountains.

He came back stunned by the gas the car didn't use. He cruised at a good rate, not printed here because, well, we don't want to rat on a pal. And the car returned 21 mpg. Not test conditions, so it can't be compared to the figures we've had for other cars, in class or out, but it's nice to have happen once in a while.

The handling, on mountain roads or test circuit, was as good and as much of a surprise.

We interrupt this road test to call Chevrolet rude names. Last year we made a big thing of a handling option code-named F-41. It came on a 427 Caprice. The feature was a rear anti-roll bar, which balanced the car on corners without hurting the ride. Great idea, we said, and everybody should go out and order one.

Somebody tried to on a 1970 Caprice, and couldn't. It is no more. Chevrolet puts the rear bar on the Chevelle SS 396, but there is no F-41 this year. Not only could we not learn why, we couldn't even persuade the factory to tell us if there had been a change. The lack of listing was obvious, but nobody knew what the only suspension option listed, by name of F-40, consisted of. We went to a dealer's showroom and looked. No rear bar. No locating devices for the rear axle except staggered shocks and multi-leaf springs. The weight distribution isn't very good; the engine is light and the car is heavy, but most of the weight is concentrated in front.

And there are now computers picking spring rates for Chevrolets. A programmer somewhere was given sets of springs, and told to allow for engines, transmissions, accessories and so on. He typed it all onto an unmutilated card. When a car is ordered, the computer reads the engine, etc., and the card, and delivers springs.

Somebody has been tampering with the machine, pouring sporting blood on it. That cold, unblinking, clattering box of tubes and wires has given the Nova SS ride rates that coddle the occupants most of the time and entertain the driver when he's driving.

The big tires help, of course. The thinking here must have been the selection of the widest tread that can go inside the Nova's fender wells, then the widest rim that fits said tire.

The Nova has a relatively low polar moment of inertia, meaning in simple terms that its major masses aren't unduly ahead or aft of the center of gravity. It's a long way from a racing car, but the theory is the same; it's easy to make the car turn, and easy to straighten it out. The Nova has one less inch of wheelbase than does the Chevelle, but total length is eight inches less. That much less overhang is that much less weight flinging itself in the wrong direction.

CAR LIFE

The springs don't have other jobs to do. The Camaro with 396 used the same suspension system, but the springs couldn't control the torque. The Dodge Challenger controls torque with massive springs, and the ride is harsh. The 300/350 doesn't have torque to control, or what torque it does have is kept in place by the staggered shocks. So the rear ride rate can be tailored to allow motion, and to match the front, i.e., just stiff enough to take some of the weight, and give the Nova a mild understeer under any normal condition.

The front tires bite when the wheel is turned. It happens right away, and they won't lose their grip unless the driver spins in positive lock to the point that the wheels are turned past optimum slip angle, something the driver is unlikely to do unless he's trying much too hard. In a fast, sweeping turn, the car stays in shape all the way through, with almost no correction required, because the Nova's cornering attitude remains the same all the time.

That's the important characteristic. There is no change of attitude in mid-corner. More power will load the springs, and the tail can be forced out, but it won't happen unless it's supposed to happen. It's a nice balance between the unswervable Chevelle and a whipsaw Z/28.

The disc/drum brakes are merely good. With that much tire we expected more, but the booster was overly eager and the proportioning spared the rear drums their share of the work. The left front wheel had a tendency to lock (surely a misadjustment of this one car, not a design flaw) and full braking couldn't be applied for the entire stop. The stopping distance was shorter than most, but not as short as it should have been.

The square Ponycar tag gets tied on when we compare Nova shape and size with Camaro and Chevelle. The

NO GAUGES, but the Nova has nice big warning lights and the controls are where they should be. Nice seats, too.

UPRIGHT STYLING makes for good rear head room. With the front seats moved back, access to the rear is limited.

BETTER THAN the Camaro is about all we can say. If it weren't for that huge tire, this space might well be useful.

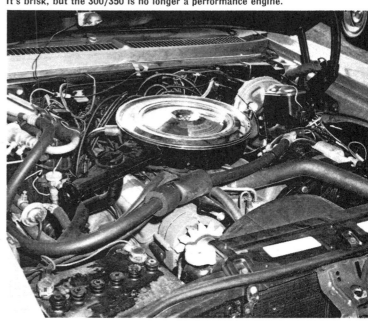

SUPERSPORT ENGINE is the 350-cid V-8 rated at 300 bhp. It's brisk, but the 300/350 is no longer a performance engine.

JANUARY 1970

NOVA

Camaro is lower and shorter, bumper to bumper and wheel to wheel. It is also much smaller inside, to the point that adults who spend any time in the rear seat are bent out of shape. The Chevelle has much more room inside.

So does Nova. Bucket front seats in both the SuperSports, and neither one really is designed to carry three people in back. But on the basis of head, shoulder and leg room, the Nova comes out a fraction in front, as if the same interior is used in both cars, with Chevelle having more nose and tail added later.

On a comfort index, Nova equals Chevelle. As a road car, provided the same engine, transmission, etc., Nova equals Camaro. Swap the sloping roofline for upright rear passengers, and the trendy short rear deck for a trunk that holds things, and you have a square Ponycar. We old fogies even liked having vent windows for the extraction of stale air.

It'll never sell. Look at the record. The Compacts were economy cars, converted with performance, comfort and trim options as soon as the will of the people was clear. (The originals didn't grow so much as they gained equipment.) The Ponycars stopped that trend. The survivors couldn't hold onto the market. When Ford had the Maverick, an economy car with eye appeal, Falcon was crushed between it and Mustang. AMC has Javelin and now Hornet, so the plain/fancy Rambler/American is gone. Plymouth always considered the Barracuda the ultimate Valiant, and that's ended, too. Boy racers that we are, we can hardly wait to test the Duster 340, but it's not aimed at the family man with enthusiast thoughts. Dodge kept the GTS going last year, but it's dropped for 1970. A man who wanted one asked about it, and we asked the factory. "We have a Ponycar now," they said, "and it's priced where the GTS used to be. And the GTS wasn't what you'd call a hot cake. Last year we sold 4,900 of 'em, which we don't interpret as overwhelming demand."

So there you are. When a maker has a zoomy small car, and a Ponycar and a top-line Compact, the dealership isn't big enough for all three. The Compact gets the axe. When

PHOTOS BY GORDON CHITTENDEN

1970 CHEVROLET
NOVA SS350

DIMENSIONS
Wheelbase, in.	111
Track, f/r, in.	59.0/58.9
Overall length, in.	189.4
width	72.4
height	52.5
Front seat hip room, in.	2 x 25
shoulder room	56.5
head room	37.6
pedal-seatback, max.	39
Rear seat hip room, in.	50
shoulder room	55.3
leg room	32.6
head room	36.6
Door opening width, in.	40.0
Trunk liftover, in.	32.0

PRICES
List, FOB factory.............$2793
Equipped as tested...........$3807
Options included: air conditioning, $363; Turbo Hydra-Matic trans., $221.80; custom exterior trim, $98; power steering, $89; posi-traction axle, $42, and evaporative controls (required in Calif.) $37.

CAPACITIES
No. of passengers	5
Luggage space, cu. ft.	13.8
Fuel tank, gal.	18
Crankcase, qt.	5
Transmission/dif., pt.	5/4.25
Radiator coolant, qt.	16

CHASSIS/SUSPENSION
Frame type: Unit steel, front subframe
Front suspension type: Independent by S.l.a., coil springs, antiroll bar.
  ride rate at wheel, lb./in. ... 91
  antiroll bar dia., in. ... 0.69
Rear suspension type: Live axle, multiple leaf springs.
  ride rate at wheel, lb./in. ... 115
Steering system: Integral assist, variable ratio recirculating ball.
  overall ratio ... 19.3:1
  turns, lock to lock ... 2.7
  turning circle, ft. curb-curb ... 40.5
Curb weight, lb. ... 3515
Test weight ... 3935
Distribution (driver), %
  f/r ... 56.2/43.8

BRAKES
Type: Power assisted disc drum.
Front rotor, dia. x width, in. .. 11.0 x 1.0
Rear drum, dia. x width .... 9.5 x 2.0
  total swept area, sq. in. ... 332.4

WHEELS/TIRES
Wheel rim size ... 14 x 7 JJ
bolt no./circle dia. in. ... 5/4.75
Tires: Uniroyal Tiger Paw.
  size ... E70-14

ENGINE
Type, no. of cyl. ... V-8
Bore x stroke, in. ... 4.00 x 3.48
Displacement, cu. in. ... 350
Compression ratio ... 10.25:1
Fuel required ... Premium
Rated bhp @ rpm ... 300 @ 4800
  equivalent mph ... 114
Rated torque @ rpm ... 380 @ 3200
  equivalent mph ... 76
Carburetion: 1 x 4 Rochester.
  throttle dia., pri./sec. .. 1.38/2.25
Valve train: Overhead valves, rocker arms, pushrods, hydraulic valve lifters.
  cam timing
    deg. int./exh. ... 28-72/78-30
    duration, int./exh. ... 280/288
Exhaust system: Dual; one muffler and two resonators.
  pipe dia., exh./tail ... 2.25/2.0
Emissions Controls: engine modifications, PCV valve, evaporative storage cannister.

DRIVE TRAIN
Transmission type: 3-speed Automatic Turbo Hydra-Matic.
Gear ratio 3rd (1.00:1) ... 3.07:1
          2nd (1.52:1) ... 4.66:1
          1st (2.52:1) ... 7.73:1
1st x t.c. stall (2.10:1) ... 16.23
axle ratio ... 3.07:1

CAR LIFE

PROVOKE THE NOVA with too much positive lock, and it plows. A lighter hand on the wheel can balance the car in turns, and combine spirited driving with a comfortable ride. The computer picked the right springs for the car.

coming car XP-887 arrives, we expect it to get here over the body of the Nova SS350. Meanwhile, Nova makes a last stand as a performance car. Pretty big talk for a middle-age compact?

Fill your hands, you sons of Volkswagen! Ol' Nova, unstyled, upright and overweight, is ready. The cavalry isn't due for another eight months, and Nova is going to lose. But the old guard is going to die fighting.

Metaphorically, of course. The Nova SS350 is not a sports car, nor a racing car, nor a Ponycar. It is a family car that goes well and handles better, a good car for a driver with passengers, or a fancier of trim and options in small packages.

It's really too bad there aren't more people like that. ∎

CAR LIFE ROAD TEST

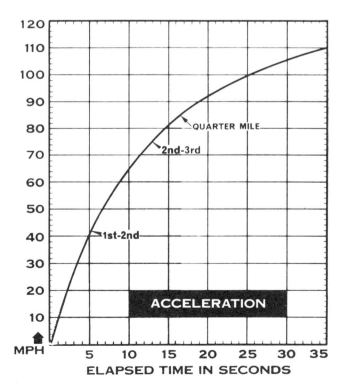

ACCELERATION

CALCULATED DATA
Lb./bhp (test weight).........13.1
Cu. ft./ton mile...............130
Mph/1000 rpm (high gear).....23.8
Engine revs/mile (60 mph).....2520
Piston travel, ft./mile........1460
CAR LIFE wear index..........36.8

SPEEDOMETER ERROR
Indicated	Actual
30 mph	25.8
40 mph	37.6
50 mph	42.2
60 mph	51.3
70 mph	59.3
80 mph	67.0
90 mph	75.0

MAINTENANCE
Engine oil, miles/days....6000/2400
  oil filter, miles/days...12,000/480
Chassis lubrication, miles.......6000
Spark plugs: ACR 44.
  gap, (in.)..............033-.038
Basic timing, deg./rpm....TDC/700
  max. cent. adv., deg./rpm.30/4700
  max. vac. adv., deg./in. Hg..20/17
Ignition point gap, in...........019
  cam dwell angle, deg........29-31
  arm tension, oz............19-23
Tappet clearance, int./exh..hudraulic
Fuel pressure at idle, psi......7.5-9
Radiator cap relief press., psi.....15

PERFORMANCE
Top speed (4900), mph..........117
Test shift points (rpm) @ mph
  2nd to 3rd (4800)..............75
  1st to 2nd (4500)..............43

ACCELERATION
0-30 mph, sec..................3.2
0-40 mph......................5.0
0-50 mph......................6.9
0-60 mph......................9.3
0-70 mph.....................11.8
0-80 mph.....................15.5
0-90 mph.....................21.0
0-100 mph....................29.8
Standing ¼-mile, sec..........16.5
  speed at end, mph...........85.0
Passing, 30-70 mph, sec........8.6

BRAKING
Max deceleration rate & stopping dist.
from 80 mph:
  rate ft./sec./ft................26
  dist. ft......................316
after 8—½ G stops from 80 mph:
  rate, ft./sec./sec..............23
  dist. ft......................377
Overall brake performance.....good

FUEL CONSUMPTION
Test cond., mpg...............13.2

JANUARY 1970

Air inlets on hood are fakes, to serving no purpose mechanical or esthetically. However, they among the few "junk items" o the Nova.

CHEVROLET NOVA 396 SS COUP

One of the compacts-that-wasn't, it offers one of the widest selections of engine options in the business.

The Nova, the sole survivor of the Chevy II series, started life as what was referred to (by Detroit) over a decade ago as the revolutionary and vital "compact car." The Ford Falcon and the Dodge Dart are two other survivors, and like the Chevy II, never were really compact. Oh, they were compacter than most of the big sleds of the 1950's, but they sure as hell weren't Volkswagens or Renaults. A wheelbase of 111 inches is long, even though NASCAR considers that a full-sized passenger car doesn't start this side of 116 inches.

The ex-Chevy II, now Nova line of compacts, does have one important thing going for it — reflected in the Nova's excellent sales record — in that there are many, many optional engines, transmissions, and styling options available for it. While the Nova is not the sporty machine that the Camaro is, it is much easier to pick and choose your options so that you end up with a personalized machine, especially in regard to its performance.

The basic body lines for 1970 are unchanged from last year, which will make most aficionados happy. Detroit has a great love of changing a design every year, even if it has reached a high level of artistic and engineering integrity. With minor exceptions, the Nova is at such a point. The simulated louvers on the front fenders, and the simulated hood air intake, add a touch of the ludicrous — but otherwise the Nova's body is very clean and sleek. The problem with these fake intakes is that they are not a part of the basic Nova, but get added on as you opt to buy the various exterior jazz-up packages, or the SS package. The least expensive is the Exterior Decor Package, at about $53 for the coupe. This includes the simulated front fender louvers and body side molding with black vinyl insert, and bright window frame molding. The Custom Exterior Package for the coupe retails for $98, and adds a ribbed rear trim panel, accent striping and back lower body accent. For the inside the Special Interior Group, at $16, will get you chrome trim on the instrument cluster, pedals, mirror support, dome light bezel, front door switch, and glove compartment light. Hot ziggetty! Top of the inside setups is the Custom Interior Package, as in our test car, with Strato-Bucket seats, special seat and sidewall trim, and all the chrome in the world, selling for $232.

Onto this you can add the Nova SS Package, which will add $291 to the tab, and includes the 300-hp 350-cu.-in. V-8 engine, dual exhausts, power front disc brakes, that awful simulated air scoop on the hood, the front fender louvers if you don't already have them (watch out they don't stick on *two* sets), black grilles fore and aft, "SS" emblems fore and aft and on the steering wheel (hey-hey!), 14×7 wheels and the E70×14 belted, bias-ply white-striped tires.

Other options that you can add at extra cost include air conditioning at $363, Positraction rear axle at $42, the

ROAD TEST

center console for $42, the special instrumentation which goes ahead of the console, and is almost impossible to scan while driving, for $95. Radios, installed at the factory, use a special antenna which is imbedded in the front window glass. Prices start with an AM pushbutton for $61, to an AM/FM radio and stereo for $239, and the ultimate — a stereo tape system with an AM/FM radio — for $373.

The 14-inch wheels can be upgraded in two steps, the Rally Wheels as on the test car adding $36 to the tab, the sexier Sport Wheels adding $79. Standard tires on the Nova line are E78×14/B. For $26, you can switch to the same tire with white stripes. Once you have the front power disc brakes, you can get the E70×14B white stripers for $44 extra, but they come as standard with the "SS" package.

Drive Train

It is in the field of power trains that the Nova offers a significant advantage over just about any other Detroit car on the market, with 8 engines and 6 transmissions, plus limited-slip and a special trailer towing rear-end ratio.

The standard engines are three. There is the little 90-hp inline 4-banger which is only offered in the Nova. In actuality it is rarely sold for this use, as it is easy to

Test Nova had 4-speed close-range manual transmission. Additional tach to left of speedometer and instrument cluster on center console are options.

Rear of Nova is uncluttered and pleasing, although fastback rear window catches dust, dew, and bird droppings constantly. It's a sacrifice for "art."

JULY 1970

Smoothness and simplicity of Nova's lines are a good reason for not making major changes every year. Fender louvers are simulated and unnecessary.

New column-lock immobilizes ignition, steering and transmission. Shifting is very sloppy and the linkage has a marked tendency to jam in two gears at once.

imagine the dreadful performance it would give in a car weighing over 3100 pounds. The standard inline-6 has 140 hp, which is bad enough. The standard V-8, and really the smallest engine one should consider even if one is the traditional Little Old Lady from Pasadena, is the 200-hp 307-cu.-in. regular gas V-8 engine. With this displacement and horsepower you are below the red-line where your auto insurance starts increasing by leaps and bounds.

If you want to play around with other engines, there is the 155-hp 250-cu.-in. inline-6, which doesn't offer much advantage over the standard six except that it is of a newer design. As we start our quest for power we have the last of the regular gasoline engines in the 250-hp 350-cu.-in. V-8. With the same displacement and premium gasoline we get 300 hp. The last two optional engines are bound to cause a great deal of confusion over the years. The "Turbo-Jet 396" comes as either 350 or 375 horsepower, the major difference being a 10.3:1 and an 11.0:1 compression ratio. However, and here is the confusing point, neither engine has a 396-cu.-in. displacement. Both are 402 cu. in. The trouble lies in the fact that Chevy has another engine which is also 402 cu. in. which they call the "Turbo-Jet 400." Added to this is the "Turbo-Fire 400" engine, which really *is* 400 cu. in., but is a "small block" engine rather than a "large block." Got that straight?

Which engine is the best? That will depend on your uses of the car, your original budget, and how much you can pay for car insurance. The Nova we tested had the 350-hp "396" engine, which seemed more than adequate for our everyday uses. It added $184 to the Nova SS Package on top of the $2700 base price of a V-8 Nova. Since the reliable and proven premium gas 300-hp 350-cu -in. V-8 is included in the "SS" Package, we would consider this a Best Buy in the Nova line if you want good performance and handling without going absolutely ape. At over 3700 pounds ready to go, the Nova isn't going to be much of a street racer, so we personally wouldn't choose to spend an additional $316 for the 375-hp "396."

With these 8 engines, you can opt to play all sorts of transmission games. There is the 3-speed Turbo Hydra-Matic in 4 models for different horsepower engines, the 2-speed Power Glide for up to 300 hp., the semi-automatic Torque Drive for the low-horsepower inline-4 and inline-6s, a 4-speed wide-range manual for the 350 and "396" engines, the 4-speed close-ratio manual for the "396" engines only, and the "standard" 3-speed manual.

One of the problems in testing any new model of automobile is trying to separate the sins of that one test car from the sins of the whole series. The 4-speed close-ratio Muncie transmission that was in our test Nova gave us endless and constant trouble. Undoubtedly much of this was the fault of the particular transmission in this particular car, but we must consider that the Muncie has a reputation for not being absolutely "top drawer" in reliability. Twice in our two-week test period we had parking lot attendants get the box jammed into two gears simultaneously. The only cure was to have the car towed to the nearest Chevy garage. The repair is relatively simple, but it's nothing the

This transmission problem presents one of the few very strong reasons we cannot give the Nova a higher rating as an overall machine. With the Turbo Hydra-Matic the problem would not exist, but the driver wouldn't have the fun of picking his own gears. Still, with the high risk of finding your car absolutely immobile everytime you get into it, we'd recommend the automatic transmission if you are set on buying a Nova. The Turbo Hydra-Matic is available for every engine except the inline-4, so there's no problem here.

Power and Performance

The 350-hp 396/402 engine in our test Nova was red-lined at 5000 rpm on the miniature tachometer alongside the huge speedometer. With all that power on the highways in third.

One place the horsepower and torque show up is in fourth gear driving. We really hadn't thought too much of the Nova's performance until we tried some acceleration runs from 40 and 50 mph in top gear. This took us right through the lumpy 2000-3000 range, which we normally avoided. We literally shot from 40-65 mph in 5.7 seconds when we floorboarded the Nova in fourth gear. A 50-65 shot took 3.2 seconds.

Roadability and Handling

The Nova is one of those automobiles about which you can't say very much either good nor bad when it comes to handling. It neither makes you shudder nor say "whoopee." It does not have the tight accuracy of a pony car, as say the Camaro; neither does it have the cold-sweat sloppiness of almost all Detroit cars a decade or so ago. In other words, it's nothing to talk about, nor is it anything to bitch about. While the suspen-

Rear leg-room is minimal in the Nova, as shown here by RT's six-foot test driver with the front seat in its normal position. Seats tip forwards.

Nova roof line is lower than it looks, as this photo with six-foot test driver shows. Special "rally" wheels are not part of the "SS" package.

owner could do in the boonies, nor is it apparently a cure — just a repair. Half a dozen times the gearbox temporarily hung up in reverse. With a bit of lever flipping and clutch popping, there would eventually be a painful grunch and the linkage problem would be temporarily solved.

Another problem of the Muncie in the test Nova, and we would assume this to be typical, is that the gear lever positions are sloppily vague. In the dark, or without a studied glance, it is virtually impossible to tell whether one is starting in first or third gear. The two slots are very close together, with no spring loading to differentiate them. At night it evolves into a practice of slipping the clutch and seeing if the engine starts to stall. If it does, you're in third: move the lever slightly to the left and try again.

JULY 1970

and 415 pound-feet of torque, fourth gear is almost useless for anything but really fast highway cruising. Taking the Nova up to the red-line in second gear will net you 70 mph, for Crying Out Loud! In fourth the engine is only turning over at 2000 rpm at 55 mph, and 3000 rpm at 65. This presented one other problem in our test car, which is more a problem of the anti-pollution devices than of the car, we feel. The engine ran very lumpily in the 2000-3000 rpm range. Unfortunately this is right where you are in third gear around town, and at 60-65 mph on the freeways. If it hurts your sensibilities to hear an engine running badly, the only choice is to drop back a gear to raise the revs 1000 or so. Thus, we ended up driving on city streets in second, with more exhaust noise than we would prefer, and

sion on American cars has improved over the years, a great deal of credit for improved handling must go to the use of the big, fat wide-oval tires that are standard on so many higher-performance cars today. All Novas come with "fiberglass bias-belted" tires.

Brakes and Safety

The power-assisted front disc brakes are not standard throughout the Nova line, as front discs are on the Camaros, but do come with the "SS" Package or can be added for $64. Particularly with a coupe or sedan weighing well over a ton and a half, we would certainly recommend them. Again, the reason is not that they are inherently better in simply stopping the car, but that they will do so repeatedly due to their lack of heat-fade

85

Body styling for the 1970 Chevrolet Nova is virtually unchanged from last year, with minor changes to the front grille and the tail lights.

Trunk offers close to 14 cubic feet of storage space, although big E70x14B/2 spare takes up quite a bit of it.

or wetness-fade. If this doesn't impress you, you might consider that you can't run the Sports or Rally Wheels unless you have the front discs!

Chevy, as every other car-maker, likes to plug its safety equipment as though it were something exclusively and originally its alone. This is nonsense, of course, as can be easily verified by checking Chevies — or any other American car — made in the years before the Federal Government required this equipment. All in all, automobile makers have about as much sincere interest in automotive safety, anti-polluting engines, and economic transportation as munition makers have in peace in our times.

To put it another way, the Nova has the same safety equipment you'll find in every other American car on today's market. Knobs and handles are recessed and/or padded. The steering column is not too apt to skewer you in a collision. The seat belt is vaguely adequate; the shoulder belt is impossible. The car stops in a reasonably straight line on dry pavement, if you don't try a panic stop; a rain-slicked highway could be another question.

Incidentally, but perhaps of limited importance, the Chevy Nova SS rated only fifth best out of six entries in the Union 76 Performance braking tests at Daytona. The Mustang Mach 1 was worse, the Plymouth Barracuda, the Dodge Challenger 340, the Javelin SST, and the Mercury Cougar XR-7 were better — in that order. With the 300-hp 350-cu.-in. V-8, the Nova SS *did* get the best gas mileage (17.503 mpg), but was sixth out of six on acceleration times.

The 200-hp 307-cu.-in. "standard V-8" Nova with drum brakes did a little better percentagewise, rating third best on braking, fifth best on acceleration, and tops again on fuel economy.

Comfort and Convenience

The Nova is a comfortable car to drive, particularly with the Strato-Bucket seats that come with the $232 Custom Interior Package. With the manual transmission, the floor pedals are well-placed and present no problem. The shift lever is right at hand, even if one can't always be certain of starting in first gear.

Getting in and out of the rear section of the 2-door coupe is a bit of a chore, but then it usually is with "tudors." The front bucket seats have a back-release button cunningly hidden near the top of the seatback which takes a bit of gymnastics to reach. If the front seat is anywhere near the rear of its track, there is almost no space between the seat bottom and the rear of the door opening. What there is, is taken up by the lap belt dispenser.

Once in the back, the legroom is quite limited unless the front seat passenger or the driver has very short legs, and keeps his seat well forward. In other words, the Nova is one of those "two-plus-two's" where the second two better be children or midgets.

ROAD TEST

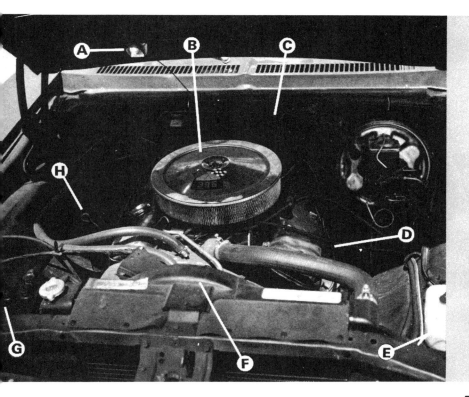

A. Engine compartment light for night service
B. Low restriction paper element air filter
C. Jumble of wires clutters engine compartment
D. Easy access to spark plugs
E. Windshield washer supply
F. Viscous drive fan with cooling shroud
G. Easy access to battery
H. Good location for dip stick

From the standpoint of the driver or the front-seat passenger, there is only the problem of the Nova's unexpectedly rough ride. Even with the wider-than-thou tires, every bump and dip in the road is transmitted to the body. Driving along a cyclically rough highway the driver finds the whole car shaking, the steering wheel vibrating, and begins to feel that the whole thing is going to come apart in his hands. The worry of "is it the road or is it the car" is a constant companion. When most modern high-performance cars are set up quite "stiff," in the competition sense, it does seem odd that the Nova — really more of a family car than a performance car — should have a markedly rougher ride than a Z-28 Camaro or a Dodge Challenger, to cite but two examples.

Economy

If nothing else, the Union 76 Performance tests showed the Novas to be the most economical in their classes (Sport Compact and Super Sport Compact) of the cars tested. The 200-hp 307-cu.-in. 2-door got 19.942 mpg, the 300-hp 350-cu.-in. got 17.503. Our testing, starting when there was slightly over 1300 miles on the engine, varied from a low of 10.0 to a high of 16.0 mpg, with an overall average of 13.0 for 662.5 miles. Naturally you'll get better mileage with the smaller-displacement engines and get poorer with the 375-hp engine. This 13.0 mpg figure, and an estimated 14.0 for the 300-hp 350-cu.-in. engine we tend to favor, comes pretty close to the national average. This is for a 3700-pound coupe, don't forget.

As to the economy of the original purchase of a Nova: The Nova 2-door coupe with the standard 307 engine and none of the interior or exterior gewgaws will run about $2650 on the price sticker on the West Coast. The price can drop by several hundreds, depending on shipping charges and how hungry the dealer is at the time you enter his showroom. The 4-door sedan costs about $30 more.

By the time you add all the options that were on our test car, you're speaking more in the range of $4200, with the same discounts for distance and hunger. If anyone were silly enough to buy the 4-cylinder coupe, it would be price-stickered at about $2500 on the West Coast.

Summary

The Nova 396 SS coupe was a decided disappointment in many ways. Although still called a compact it really isn't — and never was. It's much heavier than one would wish, scaling out at over 3100 pounds in its lightest, un-optioned form. Many of the pony cars would weigh 2700 or so in their barest form.

The array of engine options is certainly impressive, although they are for a rather unimpressive machine. The six transmissions are similarly impressive, even if the 4-speeds have all the faults that we encountered. There are so many things to play with in building up a personal and total car, that it seems a shame they're only available en masse for this particularly uninspiring model line.

Or, perhaps, that's the reason. ●

Eight engines are available for the 1970 Novas, ranging from a 90-hp inline-4 to a 375-hp big V-8. Test car had the 350-hp "396" 402-cu.-in. V-8.

Chevrolet Nova 396 SS Coupe
Data in Brief

DIMENSIONS
Overall length (in.)..................189.4
Wheelbase (in.)......................111.0
Height (in.)...........................52.5
Width (in.)............................72.4
Tread (front, in.).......................59.0
Tread (rear, in.).......................58.9
Fuel tank capacity (gal.)................(
Luggage capacity (cu. ft.)..............13.8
Turning diameter (ft.)..................40.5

ENGINE
Type................................V-8 ohc
Displacement (cu. in.) "396"...........402
Horsepower (at 5200 rpm)..............350
Torque (lb.-ft. at 3400 rpm)............415

WEIGHT, TIRES, BRAKES
Weight (curb, lb.)....................3740
Tires............................E70×14B/2
Brakes, front............11.0-in. power disc
Brakes, rear..........9.5-in. power drum

SUSPENSION
Front........independent short-and-long arm, coil springs & telescopic shock absorbers
Rear..................multi-leaf chrome-carbon steel springs

PERFORMANCE
Standing ¼ mile (sec.)................15.7
Speed at end of ¼ mile (mph)........98.02
Braking (from 60 mph, ft.)..............153

The Seven Year Car

MT Comparison

Every four or five years, U.S. manufacturers respond to a small, but vocal minority and produce an inexpensive car. And every four or five years, the lesson is relearned: cheap and inexpensive aren't necessarily the same. This time, perhaps inspired by computer printout, the cheap cars, e.g., Vega and Pinto, are not cheap. They are in fact, diminutive reproductions of their larger cousins.

Chevrolet has done it all along. The '68 Nova was designed to do a job in the small-car field and it has, every year since, quietly sold more than its quota despite the absence of significant restyling. It will remain the same until at least 1974, possibly '75, incorporating running changes into the car, even as the Vega does. If you get a late '71, then, you get a lot of '72 as well, especially if the thing comes from people who design it. So, we got three — one each from a distinct price and use class — hand delivered by the very gentlemen who monitor Nova progress — Dave McLellan, Chassis and Development; Larry Coleman, Suspension; and Jim Engle, Transmission and Exhaust.

Vega, Pinto, Gremlin, and a host of imports satisfy a need for what is, essentially, one man transport, but man is not a solo beast. He is a gregarious fellow and needs a back seat which can accommodate two adult bodies in reasonable comfort. The 1972 Nova has a fairly good sized back seat and it is reasonably comfortable. The four-door model is quite comfortable with lots of knee room for normal sized legs. The two-door has sufficient knee room but no extra space.

Chevrolet did some marketing research and came up with what may turn out to be a brilliant idea. To begin with, the researchers discovered that they sell a lot of Novas. Production figures for the 1970 model were 307,000. The strike hurt 1971 model production, with only 195,000 having been produced at this writing. In spite of the lag in numbers out the door, Nova still ranks seventh in overall nameplate production for '71.

Of the 502,000 sold over the two-year period, 89 percent were equipped with small engines, with 49 percent taking the six-cylinder 250 cid and 40 percent going with the 307 cid V8. The 350-2V went to 8 percent of the customers, but only 3 percent selected the relatively high performance 350-4V.

Although there are no customer profiles available, the significant fact that 74 percent of the Nova's sold were two-doors implies that the bulk of Nova fans are single, young marrieds or still have only small children.

The option mix for all Novas revealed a few salient factors: power steering went into 60 percent of the machines; automatic trans was bought by 80 percent of the drivers; air conditioning went in 19 percent of the units; but only 11 percent of the Nova owners selected power brakes. The arrow of optional trends points to comfort and economy.

Putting all of their data into the thinking machine, the product planners and marketeers decided to categorize their options and shoot for the relatively untouched, by Nova, performance segment of the compact arena, as well as the usual customer. Nova comes in five neatly planned well-thought-out packages for '72.

BASE MODEL or "grocery cart" version, almost totally devoid of frills. This spartan configuration is intended for use as a second car primarily in local runaround driving. Some of the suggested options are the Basic Group, consisting of wheel covers, AM radio, special interior group — day/night mirror and cigarette lighter. This can be built up from either the two-door or four-door version with the base six-cylinder engine or the eight. Power steering, power brakes and auto trans are, of course, customer's choice.

RALLY NOVA SIX, a low-priced, sporty appearing fun car. The objective of this little innovation is to provide a good handling car, with bright paint job and stripes for people who can't quite

---

Five years ago Chevrolet introduced the Nova as a cheap car. By 1974, it may be the Division's standard.

By Jim Brokaw

MOTOR TREND / SEPTEMBER 1971

Chevrolet put all of their data into the thinking machine and came out with five neatly planned, well-balanced packages for the Nova customer.

The Seven Year Car

afford a true warm dog, or don't really want one.

In addition to the basic group, the Rally Nova Six comes with a three-speed stick, sport steering wheel, and variable ratio power steering. The Rally Nova special equipment package consists of F-40 heavy-duty suspension, Rally wheels, left-hand racing mirror, stripes and floor carpet. The F-40 suspension includes a front stabilizer bar for the six-cylinder engine. The front stabilizer is standard with a V8.

Additional options provide base bucket seats, deluxe interior, power disc/drum brakes, wheel trim rings and bias-belted whitewall tires.

INSURANCE SPECIAL SPORT SEDAN, which is, in essence, a Rally Nova 8. Required equipment includes positraction limited slip differential, 350-2V engine, in addition to the Rally Nova six equipment. Deluxe buckets are the sole specific comfort option unique to the 8. This little creation is very specifically designed to counter the insurance vendetta against anything remotely resembling power.

LUXURY COMPACT is the four-door version with V8 engine. The three-speed automatic M38 transmission is a required option. Power steering, power brakes, deluxe interior and outside remote mirror are also required options.

The size, which applies logically to the entire Nova two-door line, is ideal for a commuter car, and most adequate for a small family.

MOTOR TREND / SEPTEMBER 1971

Above: Rally Nova 8 with contoured, stratobucket seats, sport steering wheel, carpeting, and floor mounted, three-speed shifter.

Below: Four-door luxury compact interior includes carpeting, sidewall trim and courtesy lighting. Air conditioning is optional extra.

Rally Nova 8, with F40 suspension, demonstrates superior cornering capacity by passing luxury sedan on the outside of turn.

Optional options include tinted glass, vinyl roof, air conditioning, AM/FM radio and bias-belted tires.

SUPER SPORT is an example of yielding to temptation. Comfort and luxury items are played down in preference to performance and handling options. Even though the 350-4V engine could hardly be called a threat to peace and security, the addition of a four-speed manual transmission will make it an instant target for most insurance companies unless your signature glows in the dark or you can walk across your swimming pool. The Super Sport package includes a 3.42:1 rear axle ratio, 7-inch wide wheels, dual exhausts and belted E70x14 tires. The F41 High Performance suspension includes a heavy-duty rear sway bar. Options also provide for "Alternative Exterior Trim Schemes." Would you believe that?

The Plain Jane "grocery cart" was omitted, assuming that the average motoring writer can mentally create his own atmosphere of boredom. Our base vehicle was the Rally Nova 6 two-door, with yellow paint and a black stripe. Steering was of the manual variety as were the drum/drum brakes. A single-barrel carburetor nursed a 250 cid six-cylinder engine. We did have the F40 heavy-duty suspension, a three-speed manual stick shift, base buckets, and a stimulating adornment of decor items.

The size, which applies logically to the entire two-door line of Novas, is ideal for a commuter car, most adequate for a two-people family or social group, fine for one or two small children, and driver's choice for anything larger. Power on the six was a disappointment. It's not a complete slug by any means, but it definitely lacks the customary low end pulling power we came to expect of sixes back when that's all Chevy had. Response throughout the usable range is uninspired, and the top end is about as exciting as a women's liberation sergeant-at-arms.

The manual steering does not require a great amount of effort to turn the small diameter sport steering wheel, but it does take an endless number of turns to get that 21:1 ratio to show anything at the front wheels. Whipping through Hollywood Boulevard traffic at 5:30 p.m., the driver is a blur from shoulders to wrists.

Manual drum/drum brakes don't make it either. Stopping distances are not unrespectable, but pedal pressure is enough to give you the leg muscles of an NFL placekicker. If you lose the challenge to feather foot the manual brakes, you may lose period. The situation is not difficult to understand but hard to justify. Chevrolet has only one basic drum brake package — with or without power assist. Since it's more of a problem to deal with the power assist, the manual-brake user has to overcome some compromise mechanical arrangement suiting both contingencies, and the end result is poorer brake reaction than we had 15 years ago.

Now, reviewing Chevrolet's own market research, you see that power brakes are specified only 11 percent of the time. We are left, then, trying to fathom why the majority of Nova buyers aren't being offered the best manual brakes, even though that is what they choose most often. What the Nova needs is a good unassisted disc front/drum rear brake arrangement.

Everything else on the six was quite up to snuff, particularly the F40 suspension, which is a joy to the heart.

All of the good pieces were on the Rally Nova 8, or insurance beater special. Quick ratio power steering, power disc/drum brakes, a 350-2V engine, fancy bucket seats, positraction rear axle, plus the normal Rally Nova equipment. Power steering converts the Nova into a nimble, quick-handling machine. Power disc/drums are sure stoppers with a capacity for minute application.

The 350-2V engine was a whole new ball game. With a torque peak of 350 lbs.-ft. at 2800 rpm, all of the go is down at the working end of the operating envelope. Acceleration is swift and sure, but not of the neck-snapping variety, as the 0-60 times of only 9.5 seconds will attest. It is, however, adequate ⟫⟫⟫

Six-cylinder, 145-hp L-6 engine gives Rally Nova adequate power, plenty of working space.

Rally Nova 8 with V8, 245 hp, 330 cid power plant gives plenty of insurance-legal power.

When air conditioning unit is added to the 350 cid, the space gets a tad crowded.

The Seven Year Car

to any task required, including freeway on-ramp operation. The three-speed manual trans is a bit of an anachronism. The primary intent of the three-speed is ostensibly economy. The only thing wrong with that theory is that in most non-cruise situations, between 20 mph and 40 mph, all of your driving is done in second gear, which puts the revs well out of the economy range. The step between first, 2.54:1, and second, 1.50:1, is just a tad steep for real economy. Fortunately, second has a very wide usable range, so that you're not constantly shifting gears, but the gas mileage is not all it would be with a four-speed.

Handling in the Rally 8 is all you could ask. Under normal circumstances, the machine is extremely stable and firm, without undue harshness. When pushed to the limit, the Nova very surprisingly goes from slight understeer to a virtual neutral condition as long as power is liberally applied. This is as you want it, inasmuch as control can be retained as long as the driver retains his cool. It is fairly difficult to spin out.

Of the luxury/comfort options, the top quality Strato-bucket seats (up to '72, available only in Camaro), are worth the price although you may not think that the rest of the fancy interior demands $247. The Strato-buckets are more generously contoured for better lateral body retention, and more heavily padded for better support for that portion of the anatomy which is most in need of cushioning.

All things considered, the Rally Nova 8, with equipment as listed on the spec page, is just about as well balanced a package as could be put together, with the exception of a four-speed shifter and the omission of good radial-ply tires instead of the E78x14 Goodyears. If you had the newest Michelins, cornering forces would only drop marginally in exchange for significantly better wet-weather performance and much improved mileage. The good people at Chevrolet tell us they would even like radials but no domestic tire manufacturer has enough capacity to service the Division's needs.

On the luxury end of things, the four-door, top-of-the-line package, with V8 engine, automatic transmission and air conditioning, is a veritable bundle of environmental control. The price is high, maybe too high, $4,066, but any car with air and a fancy interior is going to be expensive. Instead of the questionable control of the typically American cream puff boulevard ride, you get some very comfortable transportation. The standard suspension suffers from some minor front end road noise when the springs are put to the test, but it is not overly disconcerting.

There is a problem with the 350 engine when mated to the Turbo Hydramatic transmission. Seemingly because of tight emissions, on a cold start, the engine tends to stall initially when you go into drive from neutral or park. There is also a bog and surge on the first two or three cold engine accelerations from a stop. If you go to a higher idle setting, the warm engine idle will be too high, so the school solution is to keep your field boots resting on the gas pedal and hold the brake with the left foot while you engage the trans. This problem quickly dissipates as you drive along so it does not constitute a disqualifying defect.

There is an additional emission control penalty. In order to eliminate the offensive belch of pollution under compression braking, the engine is maintained at cruise rpm by means of a micro-switch in third for the manual trans, and a sensing device on the auto trans after the accelerator is released. This is good for the lungs since you won't be gassing any bystanders; however, you may suffer from involuntary expansion of the eyeballs the first couple of times you back off in third and the little machine maintains its rate of progress.

All in all, the '72 Nova comes tailor-made for your individual taste, whatever it may encompass. We feel that the ideal package is the Rally Nova 8, with power steering and brakes, four-speed manual, if it suits your fancy and your insurance agent's, or three-speed automatic if it doesn't; then, positraction, sport steering wheel, big tires and all the comfort luxury items your budget will hold. If a four-door is more suited to your needs, by all means pop for the four-door, but with the F-40 suspension and radial-ply tires. We strongly urge that you do not undertake the manual steering or manual drum/drum brakes without first road testing same.

All the signs for the next great American market swing, to the compact luxury touring car, are there for those with enough perception to see them. Chevrolet has most of the pieces in their parts bin for a head-on clash with BMW, Volvo, Audi and a host of others but don't seem to have the enthusiasm to do it.

Perhaps it is "Micro Shock." In spite of the move toward less bulk and girth in our thoughts on transportation, it is

Handling in the Rally 8 is all you could ask. Under normal circumstances the machine is extremely stable and firm, without undue harshness.

surprising to learn general American society doesn't buy small-size out of hand. While engaged in a lively discussion with a group of non-enthusiast big car drivers, one of the gentlemen described his reaction on going from a full-sized wagon, which was then under repair, to a Vega, not as a passenger, but as the driver. The sudden and dramatic shrinking of his mobile environment produced feelings not unlike that of claustrophobic panic. All of the normal judgment parameters were gone. All of the non-visual sensory inputs were alien, giving the driver no usable information. Naturally the initial response to the sub-compact experience was one of distaste. Not from actual evaluated dislike, but more from a rejection of an alien environment. Be forewarned. Even when considering a Nova, which is a bit larger than a Vega, you may have to move down from a full-sized model in steps. Chevelle first, then a Nova. Get adapted before you try to venture out behind the wheel. Naturally the younger bunch will adapt with no strain. Right?

Whatever you do in regard to Nova and its brethren, don't knock it till after you've read the price tag. /MT

6-CYL. NOVA COUPE
Base price $2,376.00
Custom deluxe belts 22.15
Soft-ray tinted glass 40.05
Rear window defroster 31.60
3-Speed floor shift 26.40
Sport steering wheel 15.80
Electric clock 16.90
AM pushbutton radio 66.40
Rally Nova equipment 99.55
Auxiliary lighting 18.45
Black cloth intereor .... NC
52-52 Sunflower Yellow (car color).... NC
Car includes front shoulder belts,
    seat back latches .... NC
                        Total $2,713.30

8-CYL. NOVA COUPE
Base price $2,471.00
Custom deluxe belts 20.55
Soft-ray tinted glass 40.05
Strato-bucket seat interior 247.65
Rear window defroster 31.60
Visor vanity mirror 3.20
Positraction rear axle 46.35
Power disc/drum brakes 69.55
245-hp Turbo-fire 350 V8 .... 26.35
3-Speed floor shift 26.40
Sport steering wheel 15.80
Power steering 103.25
E78x14 belted white stripe tires ... 54.45
Wheel trim rings 29.00
Electric clock 16.90
AM/FM pushbutton radio 139.05
Rear seat speaker 15.80
Heavy-duty radiator 14.75
Rally Nova equipment 99.55
Auxiliary lighting 18.45
Black vinyl interior .... NC
75-75 Cranberry red (car color) ..... NC
                        Total $3,489.60

8-CYL. NOVA 4-DOOR SEDAN
Base Price $2,503.00
Custom deluxe belts 22.15
Custom deluxe rear shoulder belts .. 26.35
Soft-ray tinted glass 40.05
Vinyl roof cover/green 84.30
Rear window defroster 31.60
4-Season air conditioning 391.80
Remote control rearview mirror 12.65
Visor vanity mirror 3.20
Power disc/drum brakes 69.55
245-hp Turbo-fire 350 V8 .... 26.35
Turbo Hydra-matic 205.95
Sport steering wheel 15.80
Power steering 103.25
E78x14 belted white stripe tires ... 54.45
Electric clock 16.90
AM/FM pushbutton radio 139.05
Rear seat speaker 15.80
Front and rear bumper guards 25.30
Custom interior 121.15
Custom exterior 76.90
Rally wheels 45.30
Auxiliary lighting 15.80
Black cloth interior .... NC
49-49 Antique green (car color) .... NC
Car includes: front shoulder belts; custom interior equipment — luxury trim, wood-grain door accent panels, bright interior accents, lighter, glove compartment light, carpeting, luggage mat, special insulation; custom exterior equipment — body sill and rear fender moldings, body side molding, rear trim panel .... NC
                        Total $4,046.65

SPECIFICATIONS

RALLY NOVA 6
Engine .... OHV L6
Bore & stroke — ins. .... 3.875 x 3.53
Displacement — cu. in. 250
HP @ RPM .... 145 @ 4200
Torque: lbs.-ft. @ RPM ...... 230 @ 1600
Compression Ratio/Fuel .... 8.5:1/Regular, non-leaded
Carburetion 1v
Transmission .... 3-Speed manual
Final Drive Ratio 3.08:1
Steering type .... Manual, recirculating ball nut
Steering Ratio 27.68:1
Turning Diameter (Curb-to-curb-ft.) ... 41.4
Wheel Turns (lock to lock) 4.8
Tire size E78x14
Brakes .... Manual, drum/drum
Front Suspension .... Coil/stabilizer
Rear Suspension .... Leaf spring
Body/Frame Construction .. Unit/sub-frame
Wheelbase — ins. 111
Overall length — ins. 189.4
Width — ins. 72.4
Height — ins. 52.5
Front Track — ins. 59.0
Rear Track — ins. 58.9
Curb Weight — lbs. 3,012
Fuel Capacity — gals. 16.0
Oil Capacity — qts. .... 4 (.5)
Luggage Capacity — cu. ft. 14.6

PERFORMANCE
Acceleration
  0-30 mph 4.7
  0-45 mph 7.9
  0-60 mph 12.7
  0-75 mph 21.2
Standing Start ¼-mile
  Mph 72.34
  Elapsed time 18.76
Passing speeds
  40-60 mph 6.3
  50-70 mph 8.5
Speeds in gears*
  1st ...mph @ rpm ...... 38 @ 4500
  2nd ...mph @ rpm ...... 65 @ 4500
  3rd ...mph @ rpm ...... 94 @ 4500
Mph per 1000 rpm (in top gear) 20.8
Stopping distances
  From 30 mph .... 22 ft.
  From 60 mph .... 157.9 ft.
Gas mileage range 12.9-19.6 mpg/16.58 ave.
Speedometer error
  Car
    speedometer 30 45 50 60 70 80
  Electric
    speedometer 28 45 50 61 72 82

RALLY NOVA 8
Engine .... OHV V8
Bore & stroke — ins. .... 4.00 x 3.48
Displacement — cu. in. 350
HP @ RPM .... 245 @ 4800
Torque: lbs.-ft. @ RPM ...... 350 @ 2800
Compression Ratio/Fuel .... 8.5:1/Regular, non-leaded
Carburetion 2v
Transmission .... 3-Speed manual
Final Drive Ratio 3.08:1
Steering type .... Variable ratio, power
Steering Ratjo 15.8-12.9:1
Turning Diameter (Curb-to-curb-ft.) ... 41.4
Wheel Turns (lock to lock) 3.1
Tire size E78x14
Brakes .... Power, disc/drum
Front Suspension .... Coil/stabilizer
Rear Suspension .... Leaf spring
Body/Frame Construction .. Unit/sub-frame
Wheelbase — ins. 111
Overall length — ins. 189.4
Width — ins. 72.4
Height — ins. 52.5
Front Track — ins. 59.0
Rear Track — ins. 58.9
Curb Weight — lbs. 3,250
Fuel Capacity — gals. 16.0
Oil Capacity — qts. .... 4 (.5)
Luggage Capacity — cu. ft. 14.6

PERFORMANCE
Acceleration
  0-30 mph 3.5
  0-45 mph 5.8
  0-60 mph 9.5
  0-75 mph 14.3
Standing Start ¼-mile
  Mph 81.96
  Elapsed time 16.80
Passing speeds
  40-60 mph 4.2
  50-70 mph 4.7
Speeds in gears*
  1st ...mph @ rpm ...... 54 @ 5500
  2nd ...mph @ rpm ...... 90 @ 5500
  3rd ...mph @ rpm ...... 98 @ 4000
Mph per 1000 rpm (in top gear) 24.5
Stopping distances
  From 30 mph .... 26 ft.
  From 60 mph .... 118 ft.
Gas mileage range 13.5-14.3 mpg/13.84 ave.
Speedometer error
  Car
    speedometer 30 45 50 60 70 80
  Electric
    speedometer 30 45 50 60 70 80

NOVA SEDAN V8 AUTOMATIC
Engine .... OHV V8
Bore & stroke — ins. .... 4.00 x 3.48
Displacement — cu. in. 350
HP @ RPM .... 245 @ 4800
Torque: lbs.-ft. @ RPM ...... 350 @ 2800
Compression Ratio/Fuel .... 8.5:1/Regular, non-leaded
Carburetion 2v
Transmission .... 3-Speed auto Turbo Hydra-matic
Final Drive Ratio 2.56:1
Steering type .... Variable ratio, power
Steering Ratio 15.8-12.9:1
Turning Diameter (Curb-to-curb-ft.) ... 41.4
Wheel Turns (lock to lock) 3.1
Tire size E78x14
Brakes .... Power, disc/drum
Front Suspension .... Coil/stabilizer
Rear Suspension .... Leaf spring
Body/Frame Construction .. Unit/sub-frame
Wheelbase — ins. 111
Overall length — ins. 189.4
Width — ins. 72.4
Height — ins. 53.9
Front Track — ins. 59.0
Rear Track — ins. 58.9
Curb Weight — lbs. 3,490
Fuel Capacity — gals. 16.0
Oil Capacity — qts. .... 4 (.5)
Luggage Capacity — cu. ft. 13.7

PERFORMANCE
Acceleration
  0-30 mph 3.7
  0-45 mph 6.3
  0-60 mph 9.5
  0-75 mph 14.5
Standing Start ¼-mile
  Mph 81.89
  Elapsed time 17.20
Passing speeds
  40-60 mph 5.0
  50-70 mph 5.9
Speeds in gears*
  1st ...mph @ rpm ...... 58 @ 5000
  2nd ...mph @ rpm ...... 96 @ 5000
  3rd ...mph @ rpm ..... 103 @ 4000
Mph per 1000 rpm (in top gear) 25.75
Stopping distances
  From 30 mph .... 40.5 ft.
  From 60 mph .... 146 ft.
Gas mileage range .. 13.02-15.68/14.35 ave.
Speedometer error
  Car speed-
    ometer .. 30 45 50 60 70 80
  Electric speed-
    ometer .. 31.5 46 51 61 71 80

*Speeds in gears are at shift points (limited by the length of track) and do not represent maximum speeds.

Looking toward the future, Chevrolet Engineering has developed a twin turbocharger package for the small-block V-8 engine

SUPER NOVA

TEXT AND PHOTOS BY DON GREEN ■ Yes, that's correct. A dual turbocharged Chevy Nova built entirely by Chevrolet. But we should make a couple of things clear from the very beginning. First, this Nova is in no way intended to be a race car. Dual turbochargers and all, it was designed as a street car and actually sees use on the streets of Detroit. Second, the chances of you being able to buy a similar car from Chevrolet are, at the present time, quite slim. The official position at G.M. goes something like, "There is enough interest in turbocharging that we ought to be familiar with the state of the art." They don't specifically imply that turbocharging will or even may be the way to go in the future, but they recognize the value of being well informed. The Twin Turbo is one of the results of being well informed.

The project was initiated when the people in charge of Nova development and production requested simply that something be done with turbocharging on a small-block V-8 Nova. The project was turned over to John Pierce of Chevrolet's Product Promotion Engineering, whose responsibility it became to secure the necessary hardware and follow the project through to completion, hopefully ending up with an efficient package that offers good performance and reliability while meeting the all-important Federal emissions standards.

As a starting point, a Nova was selected that was equipped with a 270 horsepower, 350 cubic-inch engine (L-48 option), Turbo-Hydro automatic transmission and standard 3.08:1 rearend. The idea was to leave the engine as stock as possible, and develop a basically bolt-on turbocharging system. By leaving the engine stock, many of the potential emissions problems would also be solved.

If you're not familiar with turbocharging, it's simply a form of supercharging—a way of pumping more fuel/air mixture into the engine's cylinders than the engine would be able to draw under normal atmospheric pressure. But instead of the pump being driven by belts or gears as a conventional supercharger, it is driven by the action of the engine's exhaust against a finned turbine wheel—just a more efficient way of driving the pump, and a method that has become almost standard on modern diesel engines.

Chevrolet built the "Twin Turbo" Chevy II to explore the possibility of turbocharging passenger cars to reduce exhaust emissions.

CAR CRAFT □ MARCH 1972

You should remember, too, that Chevrolet isn't going into this turbocharging thing blind. They've had plenty of previous production turbo experience with the *Spyder* version of the Corvair. This time, however, they took a different approach. Rather than a single turbocharger unit with the carburetor mounted on the turbo inlet side (as on the Corvair), they chose to use a pair of turbos with the carburetor on the outlet side. In effect, the turbochargers will pump air into the top of the carburetor where it will be mixed with the fuel in the conventional manner.

For the turbocharger units themselves, Chevrolet worked with the Schwitzer Corporation of Indianapolis, Indiana, one of the largest turbocharger manufacturers in the U.S. Schwitzer supplies turbocharger units for everything from International Harvester diesels to overhead-cam Indy Ford engines. Based on their experience, the people at Schwitzer were able to closely predict the requirements of the 350-inch Chevy engine and supply a pair of units that proved very compatible from the beginning of the program.

Though the job could have been handled by a single large turbocharger, two smaller units were selected for several reasons. By placing one unit on each side of the engine, close to each exhaust manifold, the amount of tubing needed to carry the exhaust to and from the turbo was greatly reduced, making the engine more accessible while making it easier to control the exhaust heat. (In turbocharging, it is desirable to keep the exhaust temperature as high as possible until the exhaust has passed through the turbine wheel housing—higher heat means a greater volume of gases and higher turbine speeds.) The two smaller turbos also eliminate the need for a waste-gate pressure relief that could become just another source of problems.

The two turbochargers supplied by Schwitzer were off-the-shelf models, normally used on the 401-cubic-inch diesel engine of a 9000-series Ford tractor. The Schwitzer model number for the turbo is 3LD220. The "3" refers to the 3-inch diameter of the turbine wheel; the "220" is the air flow—220 cubic feet per minute at a 2:1 pressure ratio.

To retain as much exhaust heat as possible, it was decided to mount the turbocharger housings directly to the cast-iron exhaust manifolds. For ease of installation, the stock Nova manifolds were replaced with the classic "ram's horn" Corvette manifolds. These were modified by having a plate welded to the front end of each manifold for the mounting of the turbo. Since the exhaust would now exit through this plate directly into the turbo, each manifold's stock outlet was plugged. Basically, that's all that is required to mount the turbocharger units. Now all that is needed is the additional plumbing. As the exhaust leaves the engine, it passes through the turbocharger, driving a similar turbine wheel that pumps air into the stock carburetor. Obviously, a second pipe must be connected to the unit to carry the exhaust to the mufflers after it leaves the turbocharger housing.

As the turbine spins, it draws air into the front section of the compressor housing, where it is compressed and delivered to the engine through a flexible hose connected to the air cleaner. At first, a stock air cleaner was used, but the 10 pounds of boost pressure that the two turbos were able to produce began to balloon the stock, stamped sheet metal air cleaner housings. Several configurations were tried, but finally a new housing was constructed from .060-inch-thick sheet stock and held together with three bolts spaced around the circumference. The new housing is fitted with an inlet from each turbo and carries a stock paper air cleaner element. The Rochester Quadra-Jet carb retains its stock jetting for emissions purposes.

The only other connection to the turbochargers

Two Schwitzer 3LD220 turbochargers fitted to the stock 4-bbl-equipped 270/350 engine with a surprising amount of clearance.

Each turbo contains a 3-inch compressor pushing 220 cfm at a 2:1 pressure ratio; restrictors hold maximum boost to 10 psi.

This innocuous opening leads air to turbo intake where turbine operating at 50-80,000 rpm boosts pressure to carb air cleaner.

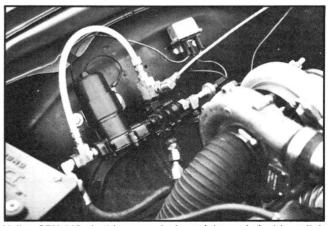

Holley GPH 110 electric pump, fuel regulator and electric switch maintain adequate fuel pressure under action of turbos' boost.

SUPER NOVA

themselves are the oil lines. The Schwitzer units are built with a floating bronze sleeve bearing on the turbine shaft, and *must* have pressure lubrication. This is supplied by flexible steel lines from the turbine housings to the stock Chevrolet oil pressure gauge connection in the front of the cylinder block. Since the turbochargers are capable of operating at speeds of up to 120,000 rpm, a ⅜-inch i.d. oil supply line is recommended. On this installation, the turbos are operated between 50,000 and 80,000 rpm, and a ¼-inch i.d. supply line is used. The line to carry the oil from the turbine housing back to the engine should have an i.d. of ¾-1 inch, since it must also carry a small amount of exhaust blowby that escapes through the snap-ring seals of the turbine housing. In the Chevy installation, the return lines carry the oil by gravity into inlets in the sides of the otherwise stock oil pan.

Braided line leading to the turbocharger supplies pressurized oil to floating bronze bearings on the exhaust turbine main shaft.

Cutaway of a typical turbine shows important parts, turbine at right is propelled by exhaust and powers compressor on left.

Chevy's turbocharger installation also makes an electric fuel pump mandatory. Since the turbos are increasing the air pressure through the carburetor, a point is reached at which there is not sufficient fuel pressure to overcome the turbos' boost pressure. As a result, fuel is unable to flow either into the carb's fuel bowls or out through the carburetor's jets. When a sensor in the air cleaner registers a boost pressure of 1 or 2 pounds, a Holley electric fuel pump is switched on. The Holley fuel pump/pressure regulator combination then automatically maintains a fuel pressure of about 5 pounds more than the boost pressure at any given time. When the engine speed is reduced and the sensor no longer registers a positive pressure in the air cleaner, the electric pump is switched off and fuel flows through a bypass. The stock mechanical fuel pump remains in operation at all times.

The whole idea of turbocharging is especially appealing on the later model cars with their lower compression ratios. The low compression helps control exhaust emissions at idle and normal cruising speed, and leaves room for the effective compression increases created by the turbocharger when the engine is operated at wider throttle positions.

We played with the Twin Turbo Nova on the test track at the Chevrolet Engineering Center in Warren, Michigan. Since the test track is not equipped with timing equipment, and since all the drag strips in the area were closed for the winter, we were limited to stopwatch timing on a measured quarter-mile distance. But this was valid enough for what we wanted—a comparison between the Twin Turbo car and a stocker. The stocker was pulled from the car pool at the Engineering Center, and promptly thundered down the quarter mile to a 16.4 at 86 mph (indicated on the speedometer). The car was a great bracket machine, since it would barely spin the tires, and produced three nearly identical runs.

The turbo car was a different story. You could have as much tire spin as you needed and often even more. With two people in the car, one manning the stop watches, we ran a best of 14.36 at an indicated 98 mph. This was with some restrictors placed in the turbocharger outlets, limiting the boost pressure to 9-10 psi. The car had previously been run without the restrictors, building a full 15 pounds of boost and turning stop-watch times of 13.90 at 103 mph. The restrictors had been placed in the turbo housings to bring the boost back to a more realistic figure to avoid possible detonation, as well as avoiding the ballooning of the new air cleaner housing. But even at 14.3, the car was more than respectable for a 350-inch Nova with a 3.08:1 rear axle.

There's nothing to match the feel of a turbocharged (or supercharged) car. The faster the engine turns, the more pressure boost the engine gets and the more horsepower it makes. About halfway through first gear, when the turbos start to get wound up, the car really begins to pull. And it doesn't feel like there's any end to the power it will develop. The Nova was pulling as strong at 100 mph as it was at 30 mph.

The real indication of how well the turbocharged Nova ran at high speeds is in the passing times, the time it takes to accelerate from a constant speed to a given higher speed. We timed both the stock Nova and the turbocharged car from 50 to 80 and 50 to 90 mph. The stocker took just under 8 seconds (7.91) to go from 50 to 70 mph, while the turbocharged car was well under 6 seconds at 5.61. Running 50 to 90 mph, the stocker dragged it out to 11.65 seconds, while the turbos cut a full 3½ seconds off that, doing it in only 8.10 seconds. That's a lot of extra time to be out in the passing lane.

At the present time, the Twin Turbo Nova is referred to as a "curiosity piece" by Chevrolet Engineering. Its sole function is to be passed around the corporation, hopefully generating some interest in turbocharging for production vehicles. We hope it does.

A Compact with Pony Car Punch

Try Chevy's Nova If You Hate Small Cars

Despite an almost disdainful attitude on the part of Chevrolet management towards updating the Nova, much less offering a bit more variety than just two body styles, that doughty if somewhat dowdy compact soldiers on as the undisputed sales leader in its field. Introduced in 1962 under the improbable name of Chevy II and intended mostly as a hedge should the public not accept the rear-engined Corvair, the car soon outsold Ford's Falcon and has successfully taken on all other newcomers to this day.

It hardly qualifies as an austerity package. Approximately 44% of its buyers order their Novas with a V8 engine, 75% specify an automatic transmission and 15% pop for air conditioning. Toss in a few more items, such as power steering and brakes that are important from both a comfort and resale standpoint, and you have a $3,400 car. You could darn near buy a couple of basic Opels for that kind of money!

We chose our test Nova with the knowledge that neither stick-shift sixes nor "SS" packages were particularly significant in the Nova sales mix. We *had* to take a 350-cubic-inch, two-barrel V8 rather than the more practical 307 because Chevrolet didn't bother to clean up the latter sufficiently to meet current California emission standards. We asked for air, power assists and an automatic. The car had a high-line interior of the bench seat variety but no exterior trim packages. It was, in essence, the average Nova right on down to the "Magic-Mirror" white paint job which is still the most popular color.

After a few days behind the wheel of a Nova, you begin to understand why a car that others might dismiss as monotonous sells so well compared to BMW 2002s, Volvo 142s and Saab 99s which are offered for the same kind of money.

The Nova demands nothing of its driver beyond the decision making involved with steering, accelerating and braking. Once the decision is made, it does what is asked of it quite competently.

The instrument panel is perhaps the best example of Chevrolet's mastery in catering to those who regard the automobile solely as a means of transportation. These people, remember, are the majority and remember too that Chevrolet is adept at the other extreme as witness the Corvette. But in any case, you have in front of you a fan-type speedometer flanked by an oversized gas gauge and an undersized clock. The usual complement of warning lights lurks in the black background, of course, but they remain anonymous and unseen until actuated by a mechanical emergency. They never "cry wolf" by appearing to glow in the daylight though you are given a bulb check each time you start the car. Their urgency is enhanced by the complete lack of other distractions on the panel. Buying the optional gauge package costs real money because it requires an extra cost console and floor shift and that, in turn, forces you into bucket seats.

Nova interiors are plain but of good quality. The single color, unrelieved by chrome or any form of plasticized veneer, does have the virtue of highlighting the items you use such as door handles and seat adjusters. However, the black vinyl in which our test car was trimmed seemed downright gloomy.

Standard tires should be replaced with oversized belted option on 7-inch rims. Koni shocks would also be a desirable addition.

The seats themselves were firm; in fact, a bit too firm for long-term comfort.

Chevrolet is now in its fourth year of experience in producing the current Nova body shells and it reflects in the tightness with which everthing fits. Were it not for the escape routes required for the flow-through ventilation system, it probably would not be possible to slam a Nova door shut with all the windows up. That idiosyncrasy used to be an exclusive of Volkswagen's who also have had a lot of practice making a single style of body. And, of course, it's a compliment to the car's builders as you can slam doors readily when the fits are sloppy.

Rear seat room in the coupe is exceptionally generous for its class. There is actually 0.3-inch more legroom in back than will be found in a Chevelle two-door and the headroom is greater by the same amount. If the front seat occupants compromise on their seat adjustments, two adults will be quite comfortable in back. In four-door sedan form, the Nova is a true five-passenger car.

The Nova uses an SLA-type front coil suspension in a stub frame that is es-

Most everything is left to the imagination except speed, gas supply and time but when a warning light does come to life, it's bright and plainly identified.

sentially similar to the Camaro design that preceded it by a year. Camaro abandoned the single leaf springs at the rear but Nova still uses them on all but cars equipped with the 350-4V engine. Both nameplates have staggered rear shock absorber mountings. So, with its Camaro inheritance, the Nova is a rather agile road car. What it loses by generally softer springing with somewhat more

ROAD TEST/FEBRUARY 1972

A 111-inch wheelbase gives the Nova coupe fractionally greater rear leg and head room than the bigger Chevelle. Retention of vent windows aids ventilation on days when air conditioning isn't needed.

travel, it gains by freedom from bottoming.

When equipped with variable ratio power steering, as was our test car, the overall ratio varies from 18.9 to 13.5, a range that suits a family-oriented car. The Camaro steering is nearly identical except for being faster. We don't think we'd want to own a Nova with unassisted steering as over five turns of the wheel are required from lock to lock. The variable ratio form of power assist is also far more precise in high-speed emergencies than a relatively sloppy manual system.

Unlike the Camaro which is nearly neutral in handling with a reasonably light engine such as the 350-V8, the Nova has fairly pronounced understeer in tight corners. We also think it's undertired in standard E78 x 14 2PR form. There's a belted option of the same size available at very little additional cost but better yet would be the choice of the optional 7-inch rim (standard is 5-inch) with suitable bigger tires. In fact, the Nova is a prime candidate for radials even at the cost of a little more harshness. Koni shocks would also be a desirable addition in lieu of the standard Delcos. V8 Novas, though, do have a stabilizer bar in front.

On concrete surfaces with rhythmic waves between joints, the Nova behaves very much like any Japanese import. It picks up and exaggerates the harmonic to the point of annoyance, although varying speed above or below the feedback range will eliminate the problem. Another point prospective owners should keep in mind is that Chevrolet's form of SLA coil suspension and associated steering linkage as used on both the Nova and the Camaro is

Even with power, pedal pressures with front discs are a bit on the high side, making it difficult to feather for maximum stops without tire damage. On plus side, Nova brake life is exceptional under average conditions.

notoriously prone to being knocked out of alignment even by moderate contact with the curb. It's worth checking right front tire wear every time you stop for gas and it can happen with the other wheel, too, if you hit a block of wood or a chunk of truck tire at speed.

Another point to remember in connection with the front suspension and steering linkage is that the Nova no longer has the so-called "permanently sealed" (actually 36,000 miles) ball joints. There now are four fittings on the suspension and seven in the steering

Ample trunk holds 14.6 cubic feet of luggage and the spare is not too hard to reach. It comes, however, devoid of any form of carpeting.

The 350-2V in the Nova's small engine compartment is a tight fit, especially with air and power assists. The car was actually designed around a six.

linkage that require greasing every four months or 6,000 miles, an interval that coincides with that recommended for oil changes. We applaud this move because it encourages a more frequent check of safety related items such as cracks on the inner sidewalls of tires, muffler condition and evidence of hydraulic fluid leakage. Also, the automatic transmission requires a fluid change every 24,000 miles, or twice as frequently if you tow a trailer.

Buyers of six-cylinder Novas who want an automatic are still forced to be contented with the obsolete two-speed Powerglide, as the semi-automatic has been dropped. With the 307 V8 in states other then California, you have a choice between Powerglide and the far better three-speed Turbo Hydra-Matic. This last is well worth its surcharge as it is as smooth as any automatic built today. More importantly, it will kick down to 2nd at speeds up to 81 mph compared to the Powerglide pooping out at 60 mph. Everyone makes a mistake in passing judgment at least once in his driving career and a box that will still downshift in the 60-75 mph range can save lives. With the Powerglide, you might not be able to finish reciting the Lord's Prayer.

It was this same feature that permitted the rather startling acceleration times we recorded at Orange County International Speedway with our test Nova. A standing quarter in 17.7 seconds at 81 mph with two-barrel carburetion and de-rated compression is only possible with a transmission that stays in second (intermediate) range, as there wasn't much left when the upshift did occur at 83 mph shortly after we passed the trap. A 0-60 time of 9.6 seconds is also quite respectable for a car that weighed 3,347 pounds in test form.

While it is doubtful if the 350-2V would ever win an economy contest (we averaged just under 15 mpg), it does stay whisper quiet throughout its normal operating range and will even chirp the tires from standstill if you rev it up a bit against the brake. Our suggestion as to ideal cruising speeds in the specification table below was made with the likes and dislikes of the highway patrol in mind. In unlimited Nevada, though, the Nova with its standard 2.73 axle would be quite happy at 90 mph or more. With kickdown flexibility being what it was, we see no real need for the optional 3.42 ratio cataloged. A locked differential, incidentally, is available for either ratio.

Another Nova feature that encourages cruising at the legal limit is its unusual freedom from wind wander. If this is true, as it was, with our test coupe, the sedan should be even more stable. We haven't heard any claim that Nova designers consulted with aerodynamicists while fashioning the body shape but in any case, they did come up with a configuration that produces superior stability. Also, a weight distribution of approximately 51/44% is not all bad for a big-engined, rear-drive compact.

Fortunately because of its side glass curvature, vent panes have been retained and we found that cracking them open a bit helped ventilation which otherwise is dependent solely on car speed. Driving with one or both front windows down and the vents partially open was also pleasant and relatively free of drafts. Heater output produced complaints of not enough from rear seat occupants unless the fan was on full blast, at which point those in front became uncomfortable. We eventually found that this imbalance was caused by the accessory floor mats blocking much of the space under the front seat and thus, air flow.

When you order front disc brakes for a surprisingly modest $70 surcharge, a Delco Moraine vacuum boost system is tossed in as a free bonus and the combination does not cost much more than specifying power assist for the standard drum brakes. Even with power, pedal pressures with the discs are a bit on the high side, making it difficult to feather for maximum stops without tire damage. The result is quite evident in our photo, but at least all four wheels lock up at precisely the same time for an unscary if uncontrolled, stop. Our stopping distance from 60 mph of 162 feet rates average. On the plus side, Nova brake life is exceptional under average conditions, one example we know of is a Nova approaching 80,000 miles without need for pad or lining replacement.

The decision as to whether to choose discs or drums depends a lot on how and where you drive. If you or a member of your family is tough on brakes, or if you commute on freeways where 60 mph traffic is spaced less than six car lengths apart, or if you spend a lot of time in the mountains, discs are a must. In lightly populated, fairly level areas, though, drums are quite adequate.